Solving Problems in Solid Mechanics
Volume 2

S. A. Urry BSc(Eng), CEng, MRAeS, FCIOB
Formerly Professor of Building Technology,
Brunel University

P. J. Turner BTech, CEng, MIMechE
Lecturer in Mechanical Engineering, Brunel
University

Longman
Scientific &
Technical

Longman Scientific & Technical,
Longman Group UK Limited,
Longman House, Burnt Mill, Harlow
Essex CM20 2JE, England
Associated companies throughout the world

First published 1986

British Library Cataloguing in Publication Data

Urry, S. A.
    Solving problems in solid mechanics.—Vol. 2.—
    (solving problems)
    1. Mechanics, Applied—Problems, exercises, etc.
    I. Title   II. Turner, P. J.
    620.1′05′076   TA350.7

ISBN 0-582-98812-8

Produced by Longman Singapore Publishers (Pte) Ltd.
Printed in Singapore.

Contents

Preface

We have written this book for students in the second and third years of engineering degree and diploma courses who are preparing for examinations in solid mechanics, stress analysis and strength of materials. The book assumes a knowledge of the basic relationships between stress, strain and elasticity, their application to simple bending and torsion, and the analysis of built-in beams, axially-loaded struts and thin shells. These topics are covered in Volume 1 but the present book can be used on its own.

Each chapter begins with a summary of theory and formulae, and this is followed by examples with detailed solutions that show how the theoretical results are used to answer numerical questions. Finally, there is a selection of practice problems with answers.

Many of the worked examples and problems have been taken from past examination papers and their sources are indicated by the following abbreviations: (UL) University of London, (Salford) University of Salford, (Brunel) Brunel University, (PCL) Polytechnic of Central London, (IMechE) Institution of Mechanical Engineers, (IStructE) Institution of Structural Engineers, and (RAeS) Royal Aeronautical Society. We are indebted to the various authorities for permission to reproduce these questions but the responsibility for the solutions is entirely ours.

We have used SI units throughout. Appendix 2 gives an explanation of the system and a table of factors for the conversion of values to and from other systems of units.

Many of the examples appeared in our earlier book *Solution of Problems in Strength of Materials and Mechanics of Solids* and we are indebted to those readers who drew our attention to errors in the previous one-volume version. Our thanks are also due to the publishers for their care during the production stages of the present book.

It is too much to hope that every error has been eliminated and any correction or comment will be gratefully acknowledged.

S. A. Urry
P. J. Turner   March 1985

1

Analysis of stress

Stresses on oblique planes

The stress on a plane perpendicular to the axis of a bar subjected to tension, compression, bending or torsion can be calculated by the methods given in Volume 1. However, a structural or machine element may fail on a plane inclined to the normal cross-section. Furthermore, it may fail with a combination of stresses each of which is insufficient to cause failure on its own. It is therefore necessary to consider the stresses on planes in all directions. In general there is, for each plane, a direct (or normal) stress, which may be tensile or compressive, and a tangential (or shear) stress.

Complementary shear stresses

A shear stress on a given plane is accompanied by another of equal magnitude on a plane perpendicular to the first.

Principal planes and stresses

In the general three-dimensional case there are three mutually perpendicular planes on which the shear stress is zero. These are called *principal planes* and the normal stresses acting on them are termed *principal stresses*. In many problems one of the principal stresses is zero and the analysis is reduced to two dimensions.

Stresses on non-principal planes

If, in a two-dimensional case, the principal stresses are σ_1 and σ_2 the normal (direct) stress σ_n and tangential (shear) stress τ on a plane making an angle θ with the principal plane on which σ_1 acts are

$$\sigma_n = \tfrac{1}{2}(\sigma_1 + \sigma_2) + \tfrac{1}{2}(\sigma_1 - \sigma_2)\cos 2\theta$$
$$\tau = -\tfrac{1}{2}(\sigma_1 - \sigma_2)\sin 2\theta$$

the minus sign in front of the expression for τ being necessary for sign convention consistency (see Example 1.2).

Mohr stress circle

Values of σ_n and τ can be determined graphically using the Mohr circle construction. Details are given in the solution to Example 1.5.

Maximum shear stress

In a two-dimensional analysis the maximum shear stress is $\tfrac{1}{2}(\sigma_1 - \sigma_2)$. If, in three dimensions, the principal stresses are σ_1, σ_2 and σ_3 and $\sigma_1 > \sigma_2 > \sigma_3$, the true maximum shear stress is $\tfrac{1}{2}(\sigma_1 - \sigma_3)$.

Combined stresses

In a two-dimensional stress system having perpendicular normal stresses σ_x and σ_y, together with complementary shear stresses τ_{xy} acting on the same planes, the normal stress σ_n in a direction making an angle θ with that of σ_x is given by

$$\sigma_n = \tfrac{1}{2}(\sigma_x + \sigma_y) + \tfrac{1}{2}(\sigma_x - \sigma_y)\cos 2\theta + \tau_{xy}\sin 2\theta$$

(the sign convention being that of Figures 1.2 and 1.3).

The corresponding shear stress τ is given by

$$\tau = \tfrac{1}{2}(\sigma_x - \sigma_y)\sin 2\theta + \tau_{xy}\cos 2\theta$$

The principal stresses σ_1 and σ_2 are the roots of the equation

$$(\sigma - \sigma_x)(\sigma - \sigma_y) = \tau_{xy}^2$$

from which

$$\sigma_1 = \tfrac{1}{2}(\sigma_x + \sigma_y) + \tfrac{1}{2}\surd[(\sigma_x - \sigma_y)^2 + 4\tau_{xy}^2]$$

and

$$\sigma_2 = \tfrac{1}{2}(\sigma_x + \sigma_y) - \tfrac{1}{2}\surd[(\sigma_x - \sigma_y)^2 + 4\tau_{xy}^2]$$

The directions of the principal stresses relative to that of σ_x are given by

$$\tan 2\theta = \frac{2\tau_{xy}}{\sigma_x - \sigma_y}$$

The maximum shear stress is

$$\tau_{max} = \tfrac{1}{2}(\sigma_1 - \sigma_2) = \tfrac{1}{2}\surd[(\sigma_x - \sigma_y)^2 + 4\tau_{xy}^2]$$

and it acts on planes which are inclined at 45° to the principal planes.

Worked examples

1.1 Stresses on an interface due to simple tension

> Deduce expressions for the normal and shear stresses on an oblique plane within a bar subjected to simple tension. Calculate the maximum shear stress on an oblique plane of a 50 mm diameter bar subjected to a pull of 100 kN. What planes have a shear stress of 16 MN/m²? What are the normal stresses on these planes?

Solution The equilibrium of a bar in tension was first mentioned in Volume 1 (Example 1.1). There, however, the stress on a cross-section perpendicular to the axis was considered. In order to establish the conditions on an oblique plane it is necessary to consider the equilibrium of a small wedge within the bar, such as ABC (Figure 1.1(a). AB represents a plane perpendicular to the axis of the bar and AC is an oblique plane (or 'interface') making an angle θ with AB.

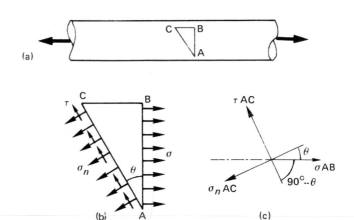

Figure 1.1

Let the axial tensile stress acting on AB (Figure 1.1(b)) be σ and suppose the triangular block ABC is kept in equilibrium by stresses σ_n and τ which are respectively normal and tangential to the plane AC.

If, for convenience, we assume the block to have unit depth perpendicular to the plane of the paper, then the *forces* acting on it are

$\sigma \times$ AB horizontally to the right,

$\sigma_n \times$ AC perpendicular to the plane AC,

$\tau \times$ AC parallel to (i.e. along) the plane AC.

These forces and the angles between them are shown in Figure 1.1(c).

For equilibrium, the sum of the resolved parts of all the forces in any direction is zero. It is usually best to resolve in directions perpendicular and parallel to the inclined plane.

Hence, resolving perpendicular to AC,

$$\sigma_n \times AC = \sigma \times AB \cos \theta$$

$$\sigma_n = \sigma \frac{AB}{AC} \cos \theta = \sigma \cos^2 \theta \qquad \text{(i)}$$

Resolving in the direction AC,

$$\tau \times AC = \sigma \times AB \cos (90° - \theta) = \sigma \times AB \sin \theta$$

$$\tau = \sigma \frac{AB}{AC} \sin \theta = \sigma \cos \theta \sin \theta = \tfrac{1}{2}\sigma \sin 2\theta \qquad \text{(ii)}$$

The expressions (i) and (ii) are independent of the size of the block ABC and, in cases where the stress σ is not uniform, can be considered as applying at a point.

The maximum value of σ_n is clearly σ itself and occurs when $\theta = 0$ (or 180°), i.e. when AC coincides with AB. Also, σ_n is zero when $\theta = 90°$.

The maximum value of τ is (from (ii)) $\frac{1}{2}\sigma$ and this occurs when $\sin 2\theta = 1$, i.e. when $2\theta = 90°$ and $\theta = 45°$.

τ is zero when $\theta = 0°$ or $90°$.

Similar results are obtained for a bar in compression.

For the bar in the question the axial stress is

$$\sigma = \frac{\text{load}}{\text{area}} = \frac{100 \text{ kN}}{\frac{1}{4}\pi \times (50 \text{ mm})^2} = 50.9 \text{ MN/m}^2 \text{ (tensile)}$$

Thus, the maximum shear stress is

$$\tau = \tfrac{1}{2}\sigma = \tfrac{1}{2} \times 50.9 \text{ MN/m}^2 = 25.45 \text{ MN/m}^2 \qquad (Ans)$$

For a shear stress of 16 MN/m² we have, from (ii),

$$\sin 2\theta = \frac{2\tau}{\sigma} = \frac{2 \times 16 \text{ MN/m}^2}{50.9 \text{ MN/m}^2} = 0.628$$

$$2\theta = 39.0° \text{ or } (180° - 39.0°) = 141.0°$$

$$\theta = 19.5° \text{ or } 70.5°$$

When $\theta = 19.5°$,

$$\text{Normal stress } \sigma_n = \sigma \cos^2 \theta$$

$$= 50.9 \text{ MN/m}^2 \times (\cos 19.5°)^2$$

$$= 45.2 \text{ MN/m}^2 \text{ (tensile)} \qquad (Ans)$$

When $\theta = 70.5°$

$$\text{Normal stress } \sigma_n = 50.9 \text{ MN/m}^2 \times (\cos 70.5°)^2$$

$$= 5.67 \text{ MN/m}^2 \text{ (tensile)} \qquad (Ans)$$

1.2 Stresses on an interface due to perpendicular direct stresses

Derive formulae for the normal and tangential stresses on an oblique plane within a material subjected to two perpendicular direct stresses.

A piece of steel plate is subjected to perpendicular stresses of 80 and 50 MN/m², both tensile. Calculate the normal and tangential stresses and the magnitude and direction of the resultant stress on the interface whose normal makes an angle of 30° with the axis of the second stress.

Solution The method is similar to that of Example 1.1. Let P_x and P_y be the forces on the plate (Figure 1.2(a)) and σ_x and σ_y the corresponding stresses. These are shown as tensile (Figure 1.2(b)) and if the sign convention of 'tensile stresses are positive, compressive are negative' is adopted then the results are applicable to all cases of direct stress.

The forces acting on the wedge ABC are shown in Figure 1.2(c) (again assuming unit depth).

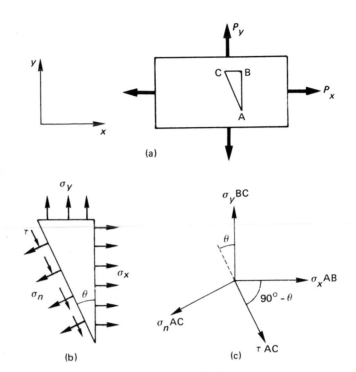

Figure 1.2

Resolving perpendicular to AC,

$$\sigma_n \times AC = \sigma_x \times AB \cos \theta + \sigma_y \times BC \cos (90° - \theta)$$

or

$$\sigma_n = \sigma_x \frac{AB}{AC} \cos \theta + \sigma_y \frac{BC}{AC} \sin \theta$$

$$= \sigma_x \cos^2 \theta + \sigma_y \sin^2 \theta$$

$$= \sigma_x \left(\frac{1 + \cos 2\theta}{2}\right) + \sigma_y \left(\frac{1 - \cos 2\theta}{2}\right)$$

$$= \tfrac{1}{2}(\sigma_x + \sigma_y) + \tfrac{1}{2}(\sigma_x - \sigma_y) \cos 2\theta \qquad\qquad \text{(i)}$$

(since $\cos 2\theta = 2 \cos^2 \theta - 1 = 1 - 2 \sin^2 \theta$).

Resolving in the direction AC,

$$\tau \times AC + \sigma_x \times AB \cos (90° - \theta) = \sigma_y \times BC \cos \theta$$

or

$$\tau = -\sigma_x \frac{AB}{AC} \sin \theta + \sigma_y \frac{BC}{AC} \cos \theta$$

$$= -\sigma_x \cos \theta \sin \theta + \sigma_y \sin \theta \cos \theta$$

$$= -(\sigma_x - \sigma_y) \sin \theta \cos \theta$$

$$= -\tfrac{1}{2}(\sigma_x - \sigma_y) \sin 2\theta \qquad\qquad \text{(ii)}$$

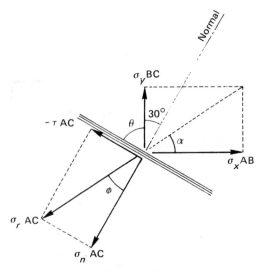

Figure 1.3

Figure 1.3 shows the relationship between the forces on the block ABC for the interface given in the problem.

The angle between the normal of the interface and the axis of σ_y is 30°. Hence

$$\theta = 90° - 30° = 60°$$

$$\sigma_x = 80 \text{ MN/m}^2 \quad \text{and} \quad \sigma_y = 50 \text{ MN/m}^2$$

$$\tfrac{1}{2}(\sigma_x + \sigma_y) = 65 \text{ MN/m}^2 \quad \text{and} \quad \tfrac{1}{2}(\sigma_x - \sigma_y) = 15 \text{ MN/m}^2$$

Using formulae (i) and (ii), the normal and tangential stresses on the interface are

$$\sigma_n = \tfrac{1}{2}(\sigma_x + \sigma_y) + \tfrac{1}{2}(\sigma_x - \sigma_y) \cos 2\theta$$
$$= 65 + (15 \times \cos 120°) = 57.5 \text{ MN/m}^2 \qquad (Ans)$$

and

$$\tau = -\tfrac{1}{2}(\sigma_x - \sigma_y) \sin 2\theta$$
$$= -15 \times \sin 120° = -13 \text{ MN/m}^2 \qquad (Ans)$$

Since the stresses σ_n and τ apply to the same area (AC) then their resultant σ_r (also based on AC) is, from Figure 1.3,

$$\sigma_r = \sqrt{[\sigma_n^2 + \tau^2]}$$
$$= \sqrt{[57.5^2 + (-13)^2]}$$
$$= 58.95 \text{ MN/m}^2 \qquad (Ans)$$

The angle made by this resultant with the normal is given by

$$\tan \phi = \tau / \sigma_n = -13/57.5 = -0.226$$
$$\phi = -12.7°$$

Also, since the three *forces* $\sigma_r \times AC$, $\sigma_x \times AB$ and $\sigma_y \times BC$ must together be in equilibrium, then (again using Figure 1.3) the angle made by the resultant with the axis of σ_x is α, where

$$\tan \alpha = \frac{\sigma_y \times BC}{\sigma_x \times AB} = \frac{\sigma_y}{\sigma_x} \tan \theta \qquad \text{(iii)}$$

With the values given,

$$\tan \alpha = \frac{50}{80} \times \tan 60° = 1.083$$

$$\alpha = 47.3° \qquad \qquad (Ans)$$

As a check, $\alpha - \phi = \theta$ (from the geometry of Figure 1.3) and, using the results obtained in the present example,

$$\alpha - \phi = 47.3° - (-12.7°) = 60°$$

1.3 Stresses on an interface due to complementary shear stresses

Explain what is meant by the term 'complementary shear stresses' and derive expressions for the normal and tangential stresses on an interface of an element in pure shear.

An element is subjected to an applied shear stress of 46 MN/m². What are the numerical values of the normal and tangential stresses on planes making angles of 15° and 65° respectively with the directions of the applied stresses?

Solution In this and later examples it is convenient to adopt the following sign conventions. Perpendicular axes x and y are taken with the positive directions being those of Figure 1.4(a). Normal stresses are positive when tensile, i.e. when they are directed outwards from the planes on which they act. A shear stress is positive if it and the outward normal to the plane on which it acts have the same signs in respect to the positive directions of the x and y coordinate axes. These definitions are used in the following analysis.

Consider a rectangular block PQRS (Figure 1.4(b)) so small that the stresses on its faces may be considered uniform. A shear stress is applied to the faces PQ and RS. The outward normal to the face PQ is in the positive x-direction. The shear stress on PQ is therefore positive when acting in the positive direction of y, i.e. vertically upwards. It is denoted by τ_{xy}. (Notice that the absence of a normal stress on PQ does not affect this definition.)

The shear stresses on PQ and RS produce an anti-clockwise couple on the block. For equilibrium, therefore, there must be an equal and opposite couple. This can only be provided by shear stresses along the faces PS and RQ as shown, since uniform normal stresses cannot produce a couple. The stress which is 'called out' on the faces PS and RQ is said to be complementary to the applied stress. This is denoted

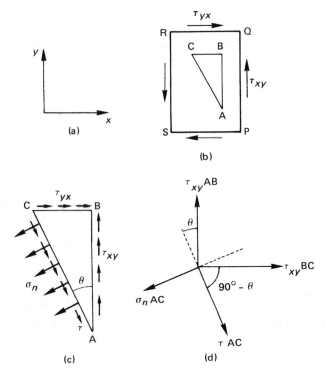

Figure 1.4

by τ_{yx} and the reader should check that the directions of the stresses shown on the faces of PQRS are all positive in accordance with the sign convention now adopted.

With unit depth of material the forces along the four faces are

$$\tau_{xy} \times PQ \qquad \tau_{yx} \times RQ \qquad \tau_{xy} \times SR \qquad \tau_{yx} \times PS$$

The couple due to the stresses τ_{xy} is $(\tau_{xy} \times PQ) \times PS$ (anti-clockwise) and that due to the stresses τ_{yx} is $(\tau_{yx} \times PS) \times PQ$ (clockwise) since PQ = SR and PS = RQ.

Equating these two couples for equilibrium

$$\tau_{xy} \times PQ \times PS = \tau_{yx} \times PS \times PQ$$

Hence

$$\tau_{xy} = \tau_{yx}$$

The stresses τ_{xy} and τ_{yx} are called complementary shear stresses. In all cases, whether there are normal stresses acting or not, the intensities of shear stresses across two planes at right angles are equal.

A material is said to be in pure (or simple) shear when it is subjected to complementary shear stresses only as in Figure 1.4(a).

The normal and tangential (shear) stresses on oblique planes are found by considering the wedge ABC, Figure 1.4(c). In accordance with the sign convention now adopted the shear stress τ on the plane AC is positive in the direction C to A. The forces acting on the wedge are shown in Figure 1.4(d).

Resolving perpendicular to AC in the usual way and replacing τ_{yx} by τ_{xy}, to which it is equal,

$$\sigma_n AC = \tau_{xy} \times BC \cos \theta + \tau_{xy} \times AB \cos (90° - \theta)$$

$$\sigma_n = \tau_{xy} \frac{BC}{AC} \cos \theta + \tau_{xy} \frac{AB}{AC} \sin \theta$$

$$= \tau_{xy} \sin \theta \cos \theta + \tau_{xy} \cos \theta \sin \theta$$

$$= 2\tau_{xy} \sin \theta \cos \theta$$

$$= \tau_{xy} \sin 2\theta \qquad\qquad\qquad\qquad\text{(i)}$$

Resolving in the direction AC

$$\tau \times AC = \tau_{xy} AB \cos \theta - \tau_{xy} BC \cos (90° - \theta)$$

$$\tau = \tau_{xy} \frac{AB}{AC} \cos \theta - \tau_{xy} \frac{BC}{AC} \sin \theta$$

$$= \tau_{xy} \cos^2 \theta - \tau_{xy} \sin^2 \theta$$

$$= \tau_{xy} \cos 2\theta \qquad\qquad\qquad\qquad\text{(ii)}$$

(since $\cos 2\theta = \cos^2 \theta - \sin^2 \theta$).

If $\tau = 46 \text{ MN/m}^2$ and $\theta = 15°$ then, using (i) and (ii),

$$\sigma_n = 46 \sin 30° = 23 \text{ MN/m}^2 \qquad\qquad (Ans)$$

$$\tau = 46 \cos 30° = 39.8 \text{ MN/m}^2 \qquad\qquad (Ans)$$

When $\theta = 65°$

$$\sigma_n = 46 \sin 130° = 35.3 \text{ MN/m}^2 \qquad\qquad (Ans)$$

$$\tau = 46 \cos 130° = -29.6 \text{ MN/m}^2 \qquad\qquad (Ans)$$

1.4 Stresses on planes inclined to the principal planes

Define *principal plane* and *principal stress*.

In a two-dimensional stress system the principal stresses are 120 MN/m² tensile and 80 MN/m² compressive. Calculate the normal, tangential and resultant stresses for planes making angles of 20° and 65° respectively with the plane of the first principal stress.

Solution A principal plane is one on which the tangential (shear) stress is zero. The normal stress on a principal plane is a principal stress. In the previous solution it was shown that the shear stresses on perpendicular planes are equal. In a two-dimensional system, therefore, there are two principal planes and these are at right angles to one another.

Let σ_1 and σ_2 be the corresponding principal stresses, also mutually perpendicular. The case of two perpendicular normal stresses was analysed in the solution to Example 1.2 of this chapter and the results given by equations (i) and (ii) of that solution can be applied to principal stresses with σ_x and σ_y being replaced by σ_1 and σ_2. Thus

$$\sigma_n = \tfrac{1}{2}(\sigma_1 + \sigma_2) + \tfrac{1}{2}(\sigma_1 - \sigma_2)\cos 2\theta$$

and

$$\tau = -\tfrac{1}{2}(\sigma_1 - \sigma_2)\sin 2\theta$$

The sign of σ_n indicates whether the normal stress is tensile or compressive; the sign of a shear stress answer is not normally of importance.

With the numerical values given in the question, and working in MN/m^2, we have $\sigma_1 = 120$ and $\sigma_2 = -80$. Substituting these figures in the results above: for $\theta = 20°$,

$$\sigma_n = \tfrac{1}{2}[120 + (-80)] + \tfrac{1}{2}[120 - (-80)]\cos(2 \times 20°)$$

$$= 96.6 \text{ MN/m}^2 \text{ tensile} \qquad (Ans)$$

$$\tau = -\tfrac{1}{2}[120 - (-80)]\sin(2 \times 20°)$$

$$= 64.3 \text{ MN/m}^2 \text{ (ignoring minus sign)} \qquad (Ans)$$

and

$$\sigma_r = \surd(96.6^2 + 64.3^2)$$

$$= 116.0 \text{ MN/m}^2 \qquad (Ans)$$

From similar calculations the corresponding results for $\theta = 65°$ are:

$$\sigma_n = 44.3 \text{ MN/m}^2 \text{ compressive}, \qquad \tau = 76.6 \text{ MN/m}^2 \quad \text{and}$$

$$\sigma_r = 88.5 \text{ MN/m}^2 \qquad (Ans)$$

1.5 Mohr stress circle

Describe the Mohr circle construction and show how it represents the stresses on an interface in a two-dimensional stress system.

Use the method to check the numerical answers to the previous question.

Solution From the previous solution the normal and tangential stresses are given by the relationships

$$\sigma_n = \tfrac{1}{2}(\sigma_1 + \sigma_2) + \tfrac{1}{2}(\sigma_1 - \sigma_2)\cos 2\theta \qquad \text{(i)}$$

and

$$\tau = -\tfrac{1}{2}(\sigma_1 - \sigma_2)\sin 2\theta \qquad \text{(ii)}$$

The Mohr circle is a geometrical construction by which these results are represented graphically. The method of finding σ_n and τ is as follows.

From a point O mark off along a horizontal axis OA and OB proportional to σ_2 and σ_1 respectively. Draw a circle on AB as diameter (Figure 1.5). For the interface required, draw a line AP making an angle θ with the horizontal. Then the horizontal and vertical coordinates of P represent the numerical values of the normal and tangential stresses on the interface.

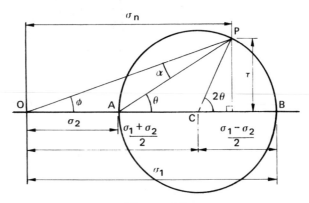

Figure 1.5

This is proved by noting that

OC $= \frac{1}{2}$(OB + OA) and thus represents $\frac{1}{2}(\sigma_1 + \sigma_2)$,
radius CB $= \frac{1}{2}$(OB − OA) and thus represents $\frac{1}{2}(\sigma_1 - \sigma_2)$,
and $\angle$PCB $= 2 \times \angle$PAB $= 2\theta$

Hence, if PN is the perpendicular from P on the horizontal axis

$$ON = OC + CN = OC + (\text{radius } CP \times \cos PCN)$$

$$= \tfrac{1}{2}(\sigma_1 + \sigma_2) + \tfrac{1}{2}(\sigma_1 - \sigma_2)\cos 2\theta$$

By comparison with (i), ON represents σ_n. Also

$$NP = \text{radius } CP \times \sin PCN$$

$$= \tfrac{1}{2}(\sigma_1 - \sigma_2)\sin 2\theta$$

By comparison with (ii), NP represents τ, except for a change in sign.

Furthermore, OP, which is the vector sum of ON and NP, represents the magnitude of the resultant stress, and its direction relative to the normal is given by ϕ (the angle POA). From the geometry of Figure 1.3 the angle made by this resultant with the axis of σ_1 is

$$\alpha = \theta - \phi$$

In Figure 1.5, angle OPA clearly equals α.

Figure 1.5 illustrates the case when σ_1 and σ_2 are both positive. If σ_2 is zero, A coincides with O. If σ_2 is negative then A is to the left of O,

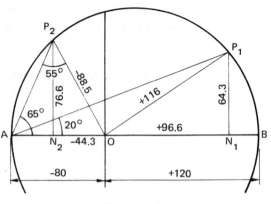

Figure 1.6

and O lies inside the circle. In this last case, normal (and resultant) stresses will be negative for points on the circle to the left of the vertical axis through O.

Figure 1.6 shows the circle for the principal stresses given in the previous question. P_1 and P_2 correspond to the 20° and 65° planes respectively. The results are stated on the lines that represent them.

1.6 Principal and maximum shear stresses due to combined direct and shear applied stresses

At a point in a bracket the stresses on two mutually perpendicular planes are 60 MN/m² tensile and 30 MN/m² tensile. The shear stress across these planes is 15 MN/m². Find, using the Mohr stress circle, the principal stresses and maximum shear stress at the point.

Solution The diagram is shown in Figure 1.7. The points F and E are marked off on the horizontal axis to represent the given direct stresses (+60 and +30 MN/m²). The shear stress is 15 MN/m² on both the planes on which these stresses act. Hence the centre of the circle C is

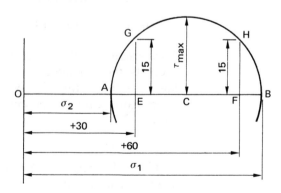

Figure 1.7

the mid-point of EF. The circle must pass through G and H where EG and FH are verticals representing the given shear stress of 15 MN/m². The circle is symmetrical about the horizontal axis so that EG and FH may be plotted upwards or downwards.

By measurement from the diagram, the principal stresses are

$$\sigma_1(=OB) = 66.2 \, MN/m^2 \qquad (Ans)$$

$$\sigma_2(=OA) = 23.8 \, MN/m^2 \qquad (Ans)$$

The maximum shear stress, which is clearly represented by the radius, is

$$\tau_{max} = 21.2 \, MN/m^2 \qquad (Ans)$$

Note The usual sign convention must be observed in the location of points E and F.

1.7 Formulae for principal stresses in two dimensions

In a two-dimensional stress system there are perpendicular direct stresses σ_x and σ_y, together with complementary shear stresses τ_{xy} on the same planes. Working from first principles, find expressions for the principal stresses.

When a certain thin-walled tube is subjected to internal pressure and torque the stresses in the tube wall are

(a) 60 MN/m² tensile,
(b) 30 MN/m² tensile in a direction at right angles to (a),
(c) complementary shear stresses of 45 MN/m² in the directions (a) and (b).

Calculate the principal stresses.

Solution Suppose, Figure 1.8, AB and BC are the planes on which the given stresses act and AC is a principal plane making an angle θ

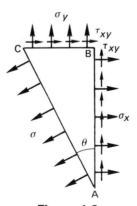

Figure 1.8

with AB. The forces acting on ABC are (considering unit depth of material):

$\sigma_x AB + \tau_{xy} BC$ horizontally to the right
$\sigma_y BC + \tau_{xy} AB$ vertically upwards
σAC perpendicular to AC (i.e. at an angle θ to the horizontal)

Resolving horizontally

$$\sigma AC \cos \theta = \sigma_x AB + \tau_{xy} BC$$

$$\sigma \cos \theta = \sigma_x \frac{AB}{AC} + \tau_{xy} \frac{BC}{AC}$$

$$= \sigma_x \cos \theta + \tau_{xy} \sin \theta$$

Hence, dividing by $\cos \theta$ and rearranging,

$$\sigma - \sigma_x = \tau_{xy} \tan \theta \tag{i}$$

Similarly, resolving vertically,

$$\sigma AC \cos (90° - \theta) = \sigma_y BC + \tau_{xy} AB$$

$$\sigma \sin \theta = \sigma_y \frac{BC}{AC} + \tau_{xy} \frac{AB}{AC}$$

$$= \sigma_y \sin \theta + \tau_{xy} \cos \theta$$

from which

$$\sigma - \sigma_y = \tau_{xy} \cot \theta \tag{ii}$$

Multiplying (i) by (ii)

$$(\sigma - \sigma_x)(\sigma - \sigma_y) = \tau_{xy}^2 \tag{iii}$$

since

$$\tan \theta \times \cot \theta = 1.$$

Removing brackets and rearranging,

$$\sigma^2 - (\sigma_x + \sigma_y)\sigma + (\sigma_x \sigma_y - \tau_{xy}^2) = 0$$

and, solving this quadratic for σ, the principal stresses are

$$\sigma = \tfrac{1}{2}\{(\sigma_x + \sigma_y) \pm \sqrt{[(\sigma_x + \sigma_y)^2 - 4(\sigma_x \sigma_y - \tau_{xy}^2)]}\}$$
$$= \tfrac{1}{2}(\sigma_x + \sigma_y) \pm \tfrac{1}{2}\sqrt{[(\sigma_x - \sigma_y)^2 + 4\tau_{xy}^2]}$$

Taking the positive and negative square roots separately, the two principal stresses are

$$\sigma_1 = \tfrac{1}{2}(\sigma_x + \sigma_y) + \tfrac{1}{2}\sqrt{[(\sigma_x - \sigma_y)^2 + 4\tau_{xy}^2]}$$

and

$$\sigma_2 = \tfrac{1}{2}(\sigma_x + \sigma_y) - \tfrac{1}{2}\sqrt{[(\sigma_x - \sigma_y)^2 + 4\tau_{xy}^2]}$$

The same results are readily obtained from the Mohr circle. Suppose, Figure 1.9, O is the origin, C is the centre of the circle, P and Q are the points on the circumference corresponding to the given stresses. Then

$$OC = \tfrac{1}{2}(\sigma_x + \sigma_y)$$

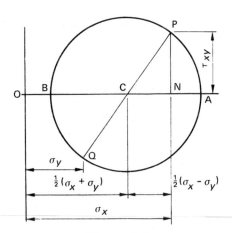

Figure 1.9

and

$$\text{radius} \quad CP = \sqrt{(CN^2 + NP^2)}$$
$$= \sqrt{\{[\tfrac{1}{2}(\sigma_x - \sigma_y)]^2 + \tau_{xy}{}^2\}}$$
$$= \tfrac{1}{2}\sqrt{[(\sigma_x - \sigma_y)^2 + 4\tau_{xy}{}^2]}$$

The principal stresses are represented by OA and OB and these distances on the diagram are given by OC ± (radius of the circle). Substitution of the dimensions determined from the diagram leads to the same relationships for σ_1 and σ_2 as those obtained earlier. This approach may not qualify as working from first principles but it provides a quick method of establishing the results.

With the numerical values given in the question, and using (i) and (ii) above,

$$\sigma_x + \sigma_y = +90 \qquad \sigma_x - \sigma_y = +30 \qquad \tau_{xy} = 45$$

Hence,

$$\sigma_1 = \tfrac{1}{2} \times 90 + \tfrac{1}{2}\sqrt{[30^2 + (4 \times 45^2)]} = 45 + 47.44$$
$$= 92.44 \text{ MN/m}^2 \text{ (tensile)} \qquad\qquad (Ans)$$

$$\sigma_2 = 45 - 47.44 = -2.44 \text{ MN/m}^2 \text{ (i.e. compressive)} \qquad (Ans)$$

1.8 Maximum principal stress in solid shaft with applied torque, bending moment and end thrust

A solid shaft 125 mm diameter transmits 600 kW at 300 rev/min. It is also subjected to a bending moment of 9 kN m and to an end thrust. If the maximum principal stress is limited to 80 MN/m², determine the permissible end thrust.

Determine the position of the plane in which the principal stress acts and draw a diagram showing the position of the plane relative to the torque and the plane of the bending moment applied to the shaft.

(UL)

Solution   The torque transmitted by the shaft is

$$T = \frac{\text{power}}{\text{angular speed}} = \frac{600 \times 10^3 \text{ W}}{300 \times 2\pi/60 \text{ rad/s}} = 19.1 \text{ kN m}$$

The maximum shear stress in the shaft (see Volume 1, Chapter 5) is

$$\tau_{xy} = \frac{Tr}{J} = \frac{T}{\frac{1}{16}\pi d^3} = \frac{16 \times 19.1 \text{ kN m}}{\pi \times (0.125 \text{ m})^3}$$

$$= 49.8 \text{ MN/m}^2$$

The direct stresses due to the bending moment and the end thrust act in the same direction. At the point where the bending stress is a maximum we have, since there is no direct stress at right angles to the bending stress,

$\sigma_x = $ max. bending stress + stress due to end thrust

$\sigma_y = 0$ and $\tau_{xy} = 49.8 \text{ MN/m}^2$

The answers may be obtained from the formulae for principal stresses derived in the previous solution or the relationship (iii) from which they were obtained. The latter is more convenient when the principal stress is given, as in the present example. Thus:

$$(\sigma - \sigma_x)(\sigma - \sigma_y) = \tau_{xy}^2$$

Putting $\sigma = -80$ (the negative sign indicating compression) in this result,

$$(-80 - \sigma_x)(-80 - 0) = 49.8^2$$

$$6400 + 80\sigma_x = 2480$$

from which

$$\sigma_x = (2480 - 6400)/80 = -49 \text{ MN/m}^2$$

the minus sign again signifying compression.
   From the bending equation (see Volume 1, Chapter 3), the maximum bending stress is

$$\frac{My_{max}}{I} = \frac{M}{\frac{1}{32}\pi d^3} = \frac{32 \times 9 \text{ kN m}}{\pi \times (0.125 \text{ m})^3} = 46.95 \text{ MN/m}^2$$

Hence the maximum permissible stress due to the end thrust is

$$(\sigma_x - \text{max. bending stress}) = (49 - 46.95) \text{ MN/m}^2$$

$$= 2.05 \text{ MN/m}^2$$

and

Permissible end thrust = stress × area

$$= 2.05 \text{ MN/m}^2 \times [\tfrac{1}{4}\pi \times (0.125 \text{ m})^2]$$

$$= 25.2 \text{ kN} \qquad\qquad (Ans)$$

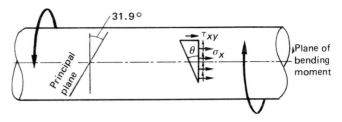

Figure 1.10

The principal plane is most easily found by using equation (i) of Example 1.7. This gives

$$\tan \theta = \frac{\sigma - \sigma_x}{\tau_{xy}} = \frac{-80 - (-49)}{49.8} = -0.6225$$

and

$$\theta = -31.9° \quad \text{or} \quad (180° - 31.9°)$$

Figure 1.10 shows the position of the principal plane relative to the axis of the bar.

The value of θ can also be found from the relationship

$$\tan 2\theta = 2\tau_{xy}/(\sigma_x - \sigma_y)$$

but this gives two results corresponding to the two principal planes.

1.9 Principal and maximum shear stresses in the web of a girder

In a two-dimensional stress system there are perpendicular direct stresses σ_x and σ_y together with complementary shear stresses τ_{xy} acting on the same planes. State, without proof, formulae for the principal stresses and maximum shear stress, and the relationships between the planes on which they act.

Discuss the special cases that arise when one of the given stresses is zero, or two of them are zero.

At a point in the web of a girder the bending stress is 80 MN/m² (tensile) and the shearing stress at the same point is 30 MN/m².

Calculate:

(a) the principal stresses,
(b) the maximum shear stress,
(c) the tensile stress which, when acting alone, would produce the same maximum shear stress.
(d) the shearing stress which, when acting alone, would produce the same maximum principal stress.

Draw a diagram showing the relationship between the principal planes and the planes of maximum shear stress.

Solution The formulae for principal stresses were derived in the solution to Example 1.7. The Mohr circle shows that the maximum shearing stress is represented by the radius of the circle and is therefore equal to half the difference between the principal stresses. Thus:

$$\tau_{max} = \tfrac{1}{2}(\sigma_1 - \sigma_2) = \tfrac{1}{2}\sqrt{[(\sigma_x - \sigma_y)^2 + 4\tau_{xy}^2]}$$

It must be emphasized that τ_{max} in the present case is the maximum shear stress in the xy plane. See the next example for the corresponding three-dimensional result.

The Mohr circle also shows that the planes of maximum shear stress are at 45° to the principal planes.

Taking the special cases in turn, and using the formulae derived in the previous solution,

If σ_y is zero, the principal stresses are

$$\sigma_1 = \tfrac{1}{2}\sigma_x + \tfrac{1}{2}\sqrt{[\sigma_x^2 + 4\tau_{xy}^2]}$$
$$\sigma_2 = \tfrac{1}{2}\sigma_x - \tfrac{1}{2}\sqrt{[\sigma_x^2 + 4\tau_{xy}^2]}$$

(i)

and the maximum shear stress is

$$\tau_{max} = \tfrac{1}{2}\sqrt{[\sigma_x^2 + 4\tau_{xy}^2]}$$

(ii)

If σ_x and σ_y are both zero, the principal stresses are

$$\sigma_1 = \tfrac{1}{2}\sqrt{(4\tau_{xy}^2)} = \tau_{xy}$$
$$\sigma_2 = -\tau_{xy}$$

(iii)

and the maximum shear stress is

$$\tau_{max} = \tfrac{1}{2}\sqrt{(4\tau_{xy}^2)} = \tau_{xy}$$

(iv)

i.e. the principal stresses are both equal to τ_{xy}, but one is tensile and the other is compressive; the principal planes bisect the planes on which the applied shear stress acts and this stress is itself the maximum shear stress.

If τ_{xy} (alone) is zero, the principal stresses are

$$\sigma_1 = \tfrac{1}{2}(\sigma_x + \sigma_y) + \tfrac{1}{2}(\sigma_x - \sigma_y) = \sigma_x$$
$$\sigma_2 = \tfrac{1}{2}(\sigma_x + \sigma_y) - \tfrac{1}{2}(\sigma_x - \sigma_y) = \sigma_y$$

(v)

and the maximum shear stress is

$$\tau_{max} = \tfrac{1}{2}(\sigma_x - \sigma_y)$$

(vi)

i.e. the applied stresses are themselves the principal stresses.

If τ_{xy} and σ_y are both zero, the principal stresses are

$$\sigma_1 = \sigma_x \quad \text{and} \quad \sigma_2 = 0$$

(vii)

and the maximum shear stress is

$$\tau_{max} = \tfrac{1}{2}\sigma_x$$

(viii)

Using the values for the girder, $\sigma_x = 80\,\text{MN/m}^2$, $\sigma_y = 0$ and $\tau_{xy} = 30\,\text{MN/m}^2$.

(a) By (i), the principal stresses are

$$\sigma_1 = (\tfrac{1}{2} \times 80) + \tfrac{1}{2}\sqrt{[80^2 + (4 \times 30^2)]} = 90\,\text{MN/m}^2 \text{ (tensile)} \qquad (Ans)$$

$$\sigma_2 = (\tfrac{1}{2} \times 80) - \tfrac{1}{2}\sqrt{[80^2 + (4 \times 30^2)]} = 10\,\text{MN/m}^2 \text{ (compressive)}$$
$$(Ans)$$

(b) By (ii), the maximum shear stress is

$$\tau_{max} = \tfrac{1}{2}\sqrt{[80^2 + (4 \times 30^2)]} = 50\,\text{MN/m}^2 \qquad (Ans)$$

(c) By (viii), the required tensile stress is

$$\sigma_x = 2\tau_{max} = 2 \times 50\,\text{MN/m}^2 = 100\,\text{MN/m}^2 \qquad (Ans)$$

(d) By (iii), the required shear stress is

$$\tau = \sigma_1 = 90\,\text{MN/m}^2 \qquad (Ans)$$

The directions of the principal planes can be found using equation (i) of Example 1.6. Thus:

$$\tan\theta_1 = \frac{\sigma_1 - \sigma_x}{\tau_{xy}} = \frac{90 - 80}{30} = 0.333$$

and

$$\theta_1 = 18.4°$$
$$\theta_2 = \theta_1 + 90° = 108.4°$$

Figure 1.11 shows the relationship between the principal planes, the planes of maximum shear stress and the plane on which σ_x acts.

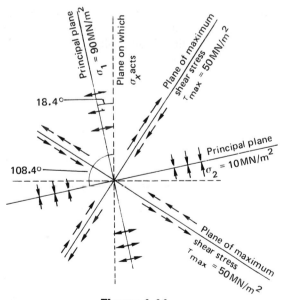

Figure 1.11

1.10 Determination of principal stresses in a three-dimensional stress system

At a point in the wall of a circular tube subjected to internal pressure and torque the stresses on three mutually perpendicular planes are as follows:

(a) A tensile stress of $120\,\text{MN/m}^2$ acting in a tangential direction.

(b) A tensile stress of $60\,\text{MN/m}^2$ acting parallel to the axis of the tube.

(c) A compressive stress of $20\,\text{MN/m}^2$ acting radially.

(d) Complementary shear stresses of $+40\,\text{MN/m}^2$ acting on the same planes as (a) and (b).

There is no shear stress on the plane on which stress (c) acts. Determine the principal stresses, the three relative maximum shear stresses and the true maximum shear stress.

Solution Consider a small element of the wall of the tube in the form of a cube as shown in Figure 1.12. Let σ_x, σ_y and σ_z be the normal stresses on such an element the stresses being positive (tensile) in the directions shown. With the sign convention adopted earlier, the shear stresses acting on the faces are positive in the directions shown and, since complementary shear stresses are equal,

$$\tau_{xy} = \tau_{yx} \qquad \tau_{yz} = \tau_{zy} \qquad \tau_{zx} = \tau_{xz}$$

Using the numerical values in the question and working in MN/m^2 throughout, we can put

$$\sigma_x = 120 \qquad \sigma_y = 60 \qquad \sigma_z = -20$$

$$\tau_{xy} = 40 \qquad \tau_{yz} = \tau_{zx} = 0$$

Since $\tau_{yz} = 0$, σ_z is a principal stress. The other principal stresses must lie within a plane at right angles to it, i.e. the plane xy. Their values

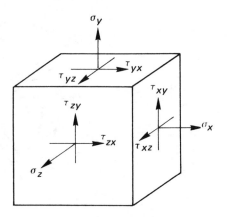

Figure 1.12

are:

$$\sigma_1 = \tfrac{1}{2}(\sigma_x + \sigma_y) + \tfrac{1}{2}\sqrt{[(\sigma_x - \sigma_y)^2 + 4\tau_{xy}^2]}$$
$$= \tfrac{1}{2}(120 + 60) + \tfrac{1}{2}\sqrt{[(120 - 60)^2 + 4 \times 40^2]}$$
$$= 90 + 50 = 140$$

and

$$\sigma_2 = 90 - 50 = 40$$

Putting $\sigma_3 = \sigma_z = -20$ the three principal stresses are $140 \, \text{MN/m}^2$ (tensile), $40 \, \text{MN/m}^2$ (tensile) and $20 \, \text{MN/m}^2$ (compressive). (*Ans*)

There are three two-dimensional stress systems in all and they are completely defined by the three principal stresses taken in pairs.

For the xy plane

$$\tau_{max} = \tfrac{1}{2}(\sigma_1 - \sigma_2) = \tfrac{1}{2}(140 - 40) = 50$$

For the yz plane

$$\tau_{max} = \tfrac{1}{2}(\sigma_2 - \sigma_3) = \tfrac{1}{2}(40 + 20) = 30$$

For the xz plane

$$\tau_{max} = \tfrac{1}{2}(\sigma_1 - \sigma_3) = \tfrac{1}{2}(140 + 20) = 80$$

The true maximum shear stress is therefore $80 \, \text{MN/m}^2$ (*Ans*)

Problems

1 A 32 mm diameter bar is subjected to a pull of 54 kN. Calculate the normal and tangential stresses on planes making angles of 5°, 37° and 79° with the axis of the bar. What is the maximum shear stress on an interface and what is the normal stress on this plane?

Answer (All stresses in MN/m^2.)
  For 5° plane, $\theta = 85°$; $\sigma_n = 0.51$; $\tau = -5.82$.
  For 37° plane, $\theta = 53°$; $\sigma_n = 24.3$; $\tau = -32.25$.
  For 79° plane, $\theta = 11°$; $\sigma_n = 64.66$; $\tau = -12.57$.
  Max. shear stress ($\theta = 45°$) is -33.55.
  Normal stress on same plane $= 33.55$.

2 Calculate the minimum diameter of a bar to carry a tensile load of 80 kN if

  (i) the tensile stress is limited to $70 \, \text{MN/m}^2$;
  (ii) the shear stress is limited to $70 \, \text{MN/m}^2$.

Answer (i) 38.2 mm, (ii) 27.0 mm.

3 In each of the following cases σ_x and σ_y are perpendicular direct stresses (in MN/m^2). Calculate the normal, tangential and resultant stresses on planes making angles of 19°, 42° and 79° with the axis of σ_x and also for the plane having the maximum shear stress. Find also the obliquity of the resultant stress (i.e. its inclination to the normal).

  (a) $\sigma_x = +60$, $\sigma_y = +15$.
  (b) $\sigma_x = +60$, $\sigma_y = -15$.
  (c) $\sigma_x = +30$, $\sigma_y = -75$.
  (d) $\sigma_x = -45$, $\sigma_y = -90$.

Answer   (All stresses in MN/m^2; ϕ is the angle of obliquity.)

  (a) For 19°, $\theta = 71°$, $\sigma_n = +19.77$; $\tau = -13.86$
        $\sigma_r = 24.09$; $\phi = -35°$

  For 42°, $\theta = 48°$, $\sigma_n = +35.14$; $\tau = -22.38$
        $\sigma_r = 41.4$; $\phi = -32.5°$

  For 79°, $\theta = 11°$, $\sigma_n = +58.36$; $\tau = -8.43$
        $\sigma_r = 58.98$; $\phi = -8.2°$

  For τ_{max}, $\theta = 45°$, $\sigma_n = +37.5$; $\tau = -22.5$
        $\sigma_r = 43.72$; $\phi = -31°$

  (b) For 19°, $\theta = 71°$, $\sigma_n = -7.05$; $\tau = -23.08$
        $\sigma_r = -24.09$; $\phi = 73°$

  For 42°, $\theta = 48°$, $\sigma_n = +18.59$; $\tau = -37.29$
        $\sigma_r = 41.67$; $\phi = -63.5°$

  For 79°, $\theta = 11°$, $\sigma_n = +57.27$; $\tau = -14.04$
        $\sigma_r = 58.98$; $\phi = -13.8°$

  For τ_{max}, $\theta = 45°$ $\sigma_n = +22.5$; $\tau = -37.5$
        $\sigma_r = 43.72$, $\phi = -59°$

  (c) For 19°, $\theta = 71°$, $\sigma_n = -63.87$; $\tau = -32.32$
        $\sigma_r = -71.56$; $\phi = 26.8°$

  For 42°, $\theta = 48°$, $\sigma_n = -27.99$; $\tau = -52.22$
        $\sigma_r = -59.2$; $\phi = 61.8°$

  For 79°, $\theta = 11°$, $\sigma_n = +26.17$; $\tau = -19.66$
        $\sigma_r = 32.74$; $\phi = -36.9°$

  For τ_{max}, $\theta = 45°$, $\sigma_n = -22.5$; $\tau = -52.5$
        $\sigma_r = -57.12$; $\phi = 66.8°$

  (d) For 19°, $\theta = 71°$, $\sigma_n = -85.23$; $\tau = -13.86$
        $\sigma_r = -86.35$; $\phi = 9.2°$

  For 42°, $\theta = 48°$, $\sigma_n = -69.85$; $\tau = -22.38$
        $\sigma_r = -73.36$; $\phi = 17.8°$

  For 79°, $\theta = 11°$, $\sigma_n = -46.65$; $\tau = -8.43$
        $\sigma_r = -47.4$; $\phi = 10.2°$

  For τ_{max}, $\theta = 45°$, $\sigma_n = -67.5$; $\tau = -22.5$
        $\sigma_r = -71.14$; $\phi = 18.3°$.

4 Calculate the principal stresses and the maximum shear stress for the following cases of two perpendicular direct stresses together with complementary shear stresses in the same directions. Draw diagrams showing the positions of the principal planes and planes of maximum shear stress relative to the planes of the applied stresses.

(a) $\sigma_x = 120$ MN/m^2 (tensile); $\sigma_y = 45$ MN/m^2 (tensile); $\tau_{xy} = +45$ MN/m^2

(b) $\sigma_x = 30$ MN/m^2 (tensile); $\sigma_y = 75$ MN/m^2 (compressive); $\tau_{xy} = +15$ MN/m^2

(c) $\sigma_x = 0$; $\sigma_y = 45$ MN/m^2 (compressive); $\tau_{xy} = -75$ MN/m^2

Answer (All stresses are in MN/m^2, positive values being tensile. θ_1 is the angle between the planes on which σ_1 and σ_x act. The angle of the 'σ_2' principal plane is $\theta_2 = \theta_1 + 90°$. Planes of maximum shear stress are 45° from the principal planes.)

(a) $\sigma_1 = +141.1$; $\sigma_2 = +23.9$; $\tau_{max} = 58.6$; $\theta_1 = 25.1°$
(b) $\sigma_1 = +32.1$; $\sigma_2 = -77.1$; $\tau_{max} = 54.6$; $\theta_1 = 7.9°$
(c) $\sigma_1 = +55.8$; $\sigma_2 = -100.8$; $\tau_{max} = 78.3$; $\theta_1 = -36.7°$

5 Solve Problem 3 (a)–(d) by the Mohr stress circle construction.

6 Draw the Mohr stress circle for the three cases given in Problem 4 and find, by measurement, the principal and maximum shearing stresses.

7 Verify the following statements from the geometry of the Mohr stress circle (Figure 1.5) σ_1 and σ_2 being the principal stresses.

(i) The maximum shear stress is $\tau_{max} = \frac{1}{2}(\sigma_1 - \sigma_2)$
(ii) The normal and resultant stresses on the plane of maximum shear stress are

$$\sigma_n = \frac{1}{2}(\sigma_1 + \sigma_2) \quad \text{and} \quad \sigma_r = \sqrt{[\frac{1}{2}(\sigma_1^2 + \sigma_2^2)]}$$

(iii) The extreme values of the resultant stress are σ_1 and σ_2
(iv) The maximum obliquity of the resultant stress is given by

$$\sin \phi = (\sigma_1 - \sigma_2)/(\sigma_1 + \sigma_2)$$

(*Hint* ϕ is a maximum when OP is a tangent.)
(v) The maximum obliquity occurs when $\theta = \phi + 45°$.

8 A solid circular shaft transmits 2240 kW at 400 rev/min and is also subjected at one section to a bending moment of 30 kN m. Calculate the minimum diameter of the shaft if the maximum shear stress is to be 60 MN/m^2.
Answer 173 mm.

9 A hollow propeller shaft, having 250 mm and 150 mm external and internal diameters respectively, transmits 1200 kW with a thrust of 400 kN. Find the speed of the shaft if the maximum principal stress is not to exceed 60 MN/m^2. What is the value of

the maximum shear stress at this speed? (*IMechE*)
Answer 80.2 rev/min; 53.63 MN/m².

10 Show that when a bending moment M and a twisting moment T act simultaneously on a circular shaft the maximum principal stress is equal to the maximum bending stress due to a simple bending moment M_E, where

$$M_E = \tfrac{1}{2}[M + \sqrt{(M^2 + T^2)}]$$

A hollow shaft is subjected to a bending moment of 2.5 kN m and a twisting moment of 3 kN m. Calculate the maximum principal stress in the shaft, the diameters being 100 mm external and 75 mm internal.
Note M_E is called the *equivalent bending moment*.
Answer 47.7 MN/m²

11 Assuming a formula for the maximum shear stress under conditions of complex stress in two dimensions, obtain an expression for an 'equivalent twisting moment' which, if acting alone, would produce the same maximum shear stress in a circular shaft as a bending moment M and twisting moment T acting together.

A solid circular shaft is to transmit 900 kW at 500 rev/min. It is supported in bearings 1.8 m apart and carries a flywheel weighing 20 kN midway between them. Find the minimum diameter of the shaft if the maximum permissible shear stress is 75 MN/m².
Answer $\sqrt{(M^2 + T^2)}$; 109.6 mm.

12 Explain the meaning of the terms 'equivalent bending moment' and 'equivalent twisting moment' in a shaft subject to both bending and torsion.

A solid shaft 76 mm diameter and spanning 3 m between bearings, has a pulley weighing 2250 N mounted at the centre of the span. The maximum speed is to be 200 rev/min when transmitting 30 kW. Find the value of the maximum principal stress in the shaft. (Neglect weight of shaft.) (*IStructE*)
Note Equivalent bending and twisting moments are discussed in Chapter 3.
Answer 45.3 MN/m².

13 Explain the meaning of principal stress. At a point in a beam there are tensile stresses of 75 and 45 MN/m² respectively at right angles to one another together with a shearing stress of 37.5 MN/m².

Calculate the maximum and minimum principal stresses in the material and the direction of the planes on which these stresses occur. (*IStructE*)
Answer 100.4 and 19.6 MN/m² (tensile) on planes making angles of 34.1° and 124.1° respectively with the plane on which the 75 MN/m² stress acts.

14 Prove that in any material in a state of shear strain a shear stress on any plane must be accompanied by a shear stress of equal intensity on any plane at right angles thereto.

At a point in the web of a beam there is a shear stress of $37.5\,\mathrm{MN/m^2}$ and a tensile stress of $90\,\mathrm{MN/m^2}$.

Find the normal and tangential components of stress on a plane making $30°$ with a plane at right angles to the direction of the tensile stress. *(IStructE)*

Answer $\sigma_n = 100.0\,\mathrm{MN/m^2}$ (tensile); $\tau = 20.2\,\mathrm{MN/m^2}$.

15 The following data apply to an electric motor driving a line shaft: output power $= 7.5\,\mathrm{kW}$; speed $= 950\,\mathrm{rev/min}$; effective diameter of motor pulley $= 120\,\mathrm{mm}$; diameter of motor shaft $= 38\,\mathrm{mm}$; distance from line of action of belt pull on driving pulley to centre of bearing $= 125\,\mathrm{mm}$, and ratio of belt tensions 2.5. For a point on the surface of the shaft at the centre of the bearing find the principal stresses and the maximum shear stress. *(IMechE)*

Answer 68.6 and $-0.72\,\mathrm{MN/m^2}$; $34.7\,\mathrm{MN/m^2}$.

16 A solid circular shaft is subjected to an axial torque T and to a bending moment M. If $M = kT$ determine in terms of k, the ratio of the maximum principal stress to the maximum shear stress. Find the power transmitted by a $50\,\mathrm{mm}$ diameter shaft, at a speed of $300\,\mathrm{rev/min}$, when $k = 0.4$ and the maximum shear stress is $75\,\mathrm{MN/m^2}$. *(IMechE)*

Answer $\text{Ratio} = \dfrac{k + \sqrt{(1 + k^2)}}{\sqrt{(1 + k^2)}}$; $53.7\,\mathrm{kW}$

17 It is calculated that the stresses on a small rectangular element of a control lever are as indicated in the diagram (Figure 1.13). Determine the principal stresses and the maximum shear stress at the element, explaining carefully your method of estimation. *(RAeS)*

Answer $45\,\mathrm{MN/m^2}$ (tensile) and $105\,\mathrm{MN/m^2}$ (compressive); $75\,\mathrm{MN/m^2}$.

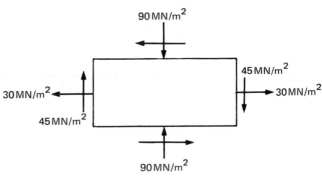

Figure 1.13

18 In a two-dimensional stress system the normal stresses at a point are σ_x and σ_y and the shear stress is τ_{xy}. Show by means of the Mohr circle diagram how the principal stresses at the point may be determined when (a) σ_x and σ_y are tensile stresses and (b) σ_x and σ_y are of opposite sign. When σ_x and σ_y are of the same sign one of the principal stresses may be zero. Illustrate this by a diagram and show how τ_{xy} is then related to σ_x and σ_y.

(*IMechE*)

Answer One principal stress is zero if $\tau_{xy}^2 = \sigma_x \sigma_y$.

19 A horizontal shaft 76 mm in diameter and 1.2 m long is held rigidly at the ends. A torque of 4.5 kN m is applied to the shaft, in a plane perpendicular to the axis of the shaft, at a distance of 0.9 m from one end. Find the magnitude of the fixing torques and the maximum shear stress in the shaft. If a concentrated load of 9 kN is now hung from the centre of the shaft find the maximum principal stress at a point on the top of the shaft, 0.15 m from the end where the lesser fixing couple occurs. (*IMechE*)

Answer Fixing torques are 3.375 kN m (at end nearest applied torque) and 1.125 kN m; maximum shear stress = 39.2 MN/m²; after load is applied required principal stress = 23.05 MN/m² (tensile).

20 In a circular shaft subjected to an axial twisting moment T and a bending moment M show that when $M = 1.2T$, the ratio of the maximum shearing stress to the greater principal stress is approximately 0.566. (*UL*)

21 The principal stresses at a point in a material are 45 MN/m² tension and 75 MN/m² tension. Working from first principles, determine, for a plane inclined at 40° to the plane on which the latter stress acts:

(a) the magnitude and angle of obliquity of the resultant stress;

(b) the normal and tangential component stresses. (*UL*)

Answer (a) $\sigma_r = 64.35$ MN/m²; $\phi = 13.3°$. (b) $\sigma_n = 62.6$ MN/m² (tensile); $\tau = 14.8$ MN/m².

22 Draw and describe Mohr's Circle of Stress and prove that it may be used to represent the state of stress at a point within a stressed material. Illustrate your answer by sketches. If, at a point within a material, the minimum and maximum principal stresses are 30 MN/m² and 90 MN/m² respectively (both tension), determine the shearing stress and normal stress on a plane passing through the point and making an angle of $\tan^{-1} 0.25$ with the plane on which the maximum principal stress acts. (*UL*)

Answer $\sigma_n = 86.5$ MN/m² (tensile); $\tau = 14.1$ MN/m².

23 At a certain point in a piece of elastic material there are normal stresses of 45 MN/m² tension and 30 MN/m² compression on two planes at right angles to one another together with shearing stresses of 22.5 MN/m² on the same planes. If the loading on the material is increased so that the stresses reach values of k times those given, find the maximum value of k if the maximum direct stress in the material is not to exceed 120 MN/m² and the maximum shearing stress is not to exceed 75 MN/m².

(UL)

Answer $k = 2.34$ for direct stress condition and 1.715 for the shearing stress condition.

24 At a certain point in a piece of material there are two planes at right angles to one another on which there are shearing stresses of 36 MN/m² together with normal stresses of 120 MN/m² tension on one plane and 56 MN/m² tension on the other plane. Determine for the given point:

(a) the magnitudes of the principal stresses,
(b) the inclinations of the principal planes,
(c) the maximum shearing stresses and the inclinations of the planes on which they act,
(d) the maximum strain if $E = 204$ GN/m² and Poisson's ratio = 0.29.

Make a diagram to show clearly the quantities and directions found in relationship to the given planes and stresses.   (UL)

Answer (a) 136.2 and 39.8 MN/m² tension;
 (b) 24.2° and 114.2° to plane on which 120 MN/m² acts;
 (c) ±48.2 MN/m² on planes at 45° to principal planes;
 (d) 0.000 611 (the method of obtaining this result is given in Chapter 2).

25 A 250 mm diameter solid shaft drives a screw propeller with an output of 6 MW. When the forward speed of the vessel is 20 knots the speed of revolution of the propeller is 240 rev/min. Find the maximum shearing stress due to the torque and the axial compressive stress due to the thrust in the shaft; hence find for a point on the surface of the shaft (a) the principal stresses and (b) the directions of the principal planes relative to the shaft axis. Make a diagram to show clearly the direction of the principal planes and stresses relative to the shaft axis.

If a graphical solution is used a clear explanation must be given of the method by which answers to both (a) and (b) were obtained. 1 knot = 0.5148 m/s.   (UL)

Answer 77.8 MN/m²; 11.9 MN/m²; (a) 83.9 MN/m² (compressive) and 72 MN/m² (tensile); (b) 42.8°, and 132.8° to the plane on which the axial compressive stress acts.

26 A 76 mm diameter steel shaft transmits a torque of 4 kN m and at the same time is subjected to an axial thrust of 44 kN. If the greatest allowable compressive stress in the shaft is 100 MN/m² find

(a) the greatest bending moment to which the shaft may be subjected in addition to the torque and thrust,

(b) the smaller principal stress and the maximum shearing stress at the point where the 100 MN/m² stress occurs.   (*UL*)

Answer   (a) 2.96 kN m. (b) 21.5 MN/m² (tensile) and 60.8 MN/m².

27 In a two-dimensional stress system (Figure 1.14) there are two planes AB and BC at right angles on which there are normal tensile stresses of 50 MN/m² and 60 MN/m² respectively together with shearing stresses of 30 MN/m². AC is any plane inclined at θ to the plane BC.

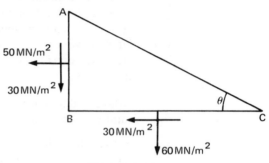

Figure 1.14

(a) Determine the normal and tangential components of the stress on AC for $\theta = 30$ degrees.

(b) Find the magnitude of θ which will make AC the principal plane on which the larger principal stress acts and determine the magnitude of this stress. Analytical or graphical methods may be used.   (*UL*)

Answer   (a) 83.5 MN/m² (tensile) and 10.67 MN/m²; (b) 40.3° and 85.4 MN/m² (tensile).

28   The stress conditions at a certain point are represented in Figure 1.15. AB and BC are two planes at right angles on which there are shearing stresses of 35 MN/m² acting towards B; there are also normal stresses of σ_x on AB and σ_y on BC. The resultant stress on AC is 75 MN/m² tension and it has an angle of obliquity of 18° as shown.

(a) Determine the magnitude of the normal stresses on AB and BC and state whether they are tension or compression.

(b) Determine the principal stresses at the point and inclinations relative to BC, of the planes on which they act.

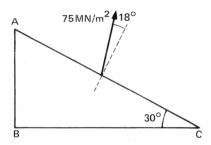

Figure 1.15

Make a diagram to show clearly the stresses and planes found.

(UL)

Answer (a) -29.5, $+64.5\,\text{MN/m}^2$; (b) $+76.1$, $-41.1\,\text{MN/m}^2$, $38.0°$, $128.0°$.

29 At a section of a rotating shaft there is a bending moment which produces a maximum direct stress of $\pm75\,\text{MN/m}^2$ and a torque which produces a maximum shearing stress of $45\,\text{MN/m}^2$.

Consider a certain point on the surface of the shaft where the bending stress is initially $75\,\text{MN/m}^2$ tension and find the principal stresses at the point in magnitude and direction:

(a) when the point is at the initial position;
(b) when the shaft has turned through $45°$;
(c) when the shaft has turned through $90°$, i.e. the point is at the neutral axis.

Make sketches to show the changes in the principal planes and stresses. *(UL)*

Answer (a) $+96.2$, $-21.2\,\text{MN/m}^2$, $25.1°$ and $115.1°$ (to axis of shaft). (b) $+78.8$, $-25.6\,\text{MN/m}^2$, $29.8°$ and $119.8°$. (c) $+45$, $-45\,\text{MN/m}^2$, $45°$ and $135°$.

30 A horizontal beam spans $10\,\text{m}$ and is simply supported at its ends. It carries a distributed load whose intensity varies linearly from zero at the left-hand end to $3\,\text{kN/m}$ at the right-hand end as shown in Figure 1.16(a). The beam is of I-section, with flanges

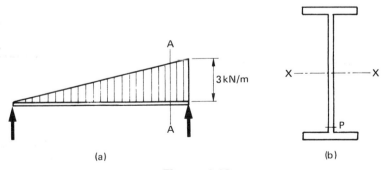

(a)   (b)

Figure 1.16

80 mm wide and 10 mm thick, a web 5 mm thick and an overall depth of 180 mm as shown in Figure 1.16(b). Take $I_{xx} = 1328 \text{ cm}^4$.

Determine for the point P, 75 mm below the neutral surface, at the section AA which is 1 m from the right-hand end:

(a) the bending stress;
(b) the vertical shear stress, assuming that the web carries the whole of the shearing force uniformly distributed;
(c) the principal stresses;
(d) the greatest shearing stress on an inclined interface;
(e) the direct stress which, if acting alone, would produce a strain equal to the greater principal strain. Take Poisson's ratio = 0.3.

Note Part (e) is covered in Chapter 2.
Answer (a) 48.3 MN/m^2 (tensile); (b) 8.94 MN/m^2; (c) 49.9 MN/m^2 (tensile) and 1.6 MN/m^2 (compressive); (d) 25.8 MN/m^2; (e) 50.4 MN/m^2 (tensile).

31 A shaft 50 mm long and 10 mm diameter carries a crank 60 mm long at right angles to the shaft axis as shown in Figure 1.17. If a vertical load of 100 N is applied to the end of the horizontal crank, sketch the Mohr's stress circle for a point A on the top of the shaft where it enters its rigid support. Calculate the maximum shear stress at A, the principal stresses and the angle the larger principal stress makes with the shaft axis.   (*Salford*)
Answer 39.8 MN/m^2; 65.24 (tensile) and 14.3 (compressive) MN/m^2; 25.1°.

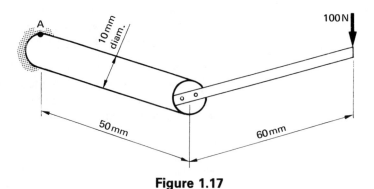

Figure 1.17

32 In a certain experiment on combined stresses, a mild steel tube, 25 mm internal diameter and 1.5 mm wall thickness, was closed at the ends and subjected to an internal fluid pressure of 840 kN/m^2. At the same time the tube was subjected to an axial pull of 886 N and to pure torsion by means of a couple, the axis of which coincided with the axis of the tube. If for the purposes

of the experiment a maximum principal stress of 36 MN/m^2 were required in the material at the outer surface of the tube, find the applied torque in N m. (IMechE)
Answer 43.3 N m.

33 A thin cylindrical tube, 76 mm internal diameter and wall thickness 5 mm, is closed at the ends and subjected to an internal pressure of 5.5 MN/m^2. A torque of 500π N m is also applied to the tube.

Determine the maximum and minimum principal stresses and also the maximum shearing stress in the wall of the tube. (UL)
Answer Principal stresses, 65.2 MN/m^2 (tensile) and 2.54 MN/m^2 (compressive); maximum shearing stress, 33.9 MN/m^2.

2

Analysis of strain

Principal strains

Under two-dimensional conditions the principal stresses (σ_1 and σ_2) and principal strains (ε_1 and ε_2) are related by

$$\varepsilon_1 = \frac{1}{E}(\sigma_1 - v\sigma_2) \quad \text{and} \quad \varepsilon_2 = \frac{1}{E}(\sigma_2 - v\sigma_1)$$

$$\sigma_1 = \frac{E}{1 - v^2}(\varepsilon_1 + v\varepsilon_2) \quad \text{and} \quad \sigma_2 = \frac{E}{1 - v^2}(\varepsilon_2 + v\varepsilon_1)$$

where E = modulus of elasticity
$\quad\quad\ v$ = Poisson's ratio

Variation of strain

The strain ε_θ in a direction making an angle θ with that of ε_1 is given by

$$\varepsilon_\theta = \tfrac{1}{2}(\varepsilon_1 + \varepsilon_2) + \tfrac{1}{2}(\varepsilon_1 - \varepsilon_2)\cos 2\theta$$

Mohr circle of strain

The form of the last result is identical to that for the normal stress on an interface, σ_n, σ_1 and σ_2 being replaced by ε_θ, ε_1 and ε_2. The Mohr circle construction can therefore be used to investigate the variation of strain with angle. It provides a useful means of analysing strain gauge rosette readings.

Volumetric strain and bulk modulus

If the volume V of a solid body changes by a small amount δV the volumetric strain is $\delta V/V$. Volumetric stress is the uniform pressure acting equally over the entire surface of the body, such as the fluid pressure acting on a body immersed in a liquid. Bulk modulus K is defined as

$$K = \frac{\text{volumetric stress}}{\text{volumetric strain}}$$

Relationships between the elastic constants

The elastic moduli E, G and K are related by

$$E = 2G(1 + v) \quad \text{and} \quad E = 3K(1 - 2v)$$

Worked examples

2.1 Determination of principal stresses from principal strains in two dimensions

A rectangular plate of steel is stressed as shown in Figure 2.1 and the strains in the X and Y directions are 11.85×10^{-5} and 9.47×10^{-5} respectively. Find the stresses σ_1 and σ_2 and hence determine the normal and shear stresses on the interface AB. ($E = 207 \text{ GN/m}^2$; $v = 0.28$.)   (*IMechE*)

Solution The strains in the directions of the principal stresses are called the *principal strains*. It is convenient to have formulae for the principal stresses in terms of these strains and, in the two-dimensional case, these are derived as follows.

Let ε_1 and ε_2 be the strains in the X and Y directions respectively. Then, if tensile stresses and strains are taken as positive,

$$\varepsilon_1 = \frac{\sigma_1}{E} - \frac{v\sigma_2}{E} \tag{i}$$

$$\varepsilon_2 = \frac{\sigma_2}{E} - \frac{v\sigma_1}{E} \tag{ii}$$

Multiplying (ii) by v we obtain

$$v\varepsilon_2 = \frac{v\sigma_2}{E} - \frac{v^2\sigma_1}{E}$$

and adding to (i),

$$\varepsilon_1 + v\varepsilon_2 = \frac{\sigma_1}{E}(1 - v^2)$$

from which

$$\sigma_1 = \frac{E}{1 - v^2}(\varepsilon_1 + v\varepsilon_2)$$

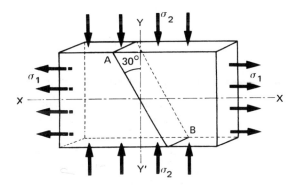

Figure 2.1

Similarly

$$\sigma_2 = \frac{E}{1 - v^2} (\varepsilon_2 + v\varepsilon_1)$$

With the numerical data given in the question

$$\varepsilon_1 = 11.85 \times 10^{-5} \quad \text{and} \quad \varepsilon_2 = -9.45 \times 10^{-5},$$

the minus sign being necessary because each of the given stresses leads to negative strains in the Y direction. Then

$$\sigma_1 = \frac{207 \times 10^9 \, \text{N/m}^2}{1 - 0.28^2} (11.85 \times 10^{-5} - 0.28 \times 9.47 \times 10^{-5})$$

$$= 20.65 \, \text{MN/m}^2 \, \text{(tensile)} \tag{Ans}$$

and

$$\sigma_2 = \frac{207 \times 10^9 \, \text{N/m}^2}{1 - 0.28^2} (-9.47 \times 10^{-5} + 0.28 \times 11.85 \times 10^{-5})$$

$$= -13.81 \, \text{MN/m}^2 \, \text{(i.e. compressive)} \tag{Ans}$$

With these results the normal and shear stresses on the given interface are

$$\sigma_\theta = \tfrac{1}{2}(\sigma_1 + \sigma_2) + \tfrac{1}{2}(\sigma_1 - \sigma_2) \cos 2\theta$$

$$= \tfrac{1}{2}(20.65 - 13.81) + \tfrac{1}{2}(20.65 + 13.81) \cos 60°$$

$$= 12.04 \, \text{MN/m}^2 \, \text{(tensile)} \tag{Ans}$$

and

$$\tau = \tfrac{1}{2}(\sigma_1 - \sigma_2) \sin 2\theta$$

$$= \tfrac{1}{2}(20.65 + 13.81) \sin 60°$$

$$= 14.92 \, \text{MN/m}^2 \tag{Ans}$$

2.2 Formulae for principal stresses in terms of principal strains

Prove that the principal stresses σ_1 and σ_2 at a point in a material can be derived from the strains ε_1 and ε_2 in the direction of the principal stresses by the equations

$$\sigma_1 = \frac{E}{1 - v^2} (\varepsilon_1 + v\varepsilon_2)$$

$$\sigma_2 = \frac{E}{1 - v^2} (\varepsilon_2 + v\varepsilon_1)$$

where v = Poisson's ratio.

In order to determine the principal stresses at a point in a structural member two strain gauges are fixed, their directions being at 30° to the known directions of the principal stresses. The measured strains in these two directions are $+455 \times 10^{-6}$ and -32×10^{-6} respectively. If $E = 200 \, \text{GN/m}^2$ and $v = 0.3$ find the magnitudes of the principal stresses. (UL)

Solution The derivation of the equations given in the question is covered in the previous solution. A note on strain gauges is given in the solution to Example 2.3.

For the second part of the question suppose (Figure 2.2) that σ_θ and $\sigma_{(\theta+90°)}$ are the normal stresses in directions each making an angle θ with the directions of the principal stresses σ_1 and σ_2. Then from the analysis given in Chapter 1,

$$\sigma_\theta = \tfrac{1}{2}(\sigma_1 + \sigma_2) + \tfrac{1}{2}(\sigma_1 - \sigma_2)\cos 2\theta$$

and

$$\sigma_{(\theta+90°)} = \tfrac{1}{2}(\sigma_1 + \sigma_2) + \tfrac{1}{2}(\sigma_1 - \sigma_2)\cos(2\theta + 180°)$$
$$= \tfrac{1}{2}(\sigma_1 + \sigma_2) - \tfrac{1}{2}(\sigma_1 - \sigma_2)\cos 2\theta$$

The strain in the direction of σ_θ is (allowing for the Poisson's ratio effect)

$$\varepsilon_\theta = \frac{\sigma_\theta}{E} - \frac{v\sigma_{(\theta+90°)}}{E}$$

$$= \frac{1}{2E}[(\sigma_1 + \sigma_2) + (\sigma_1 - \sigma_2)\cos 2\theta - v(\sigma_1 + \sigma_2)$$
$$+ v(\sigma_1 - \sigma_2)\cos 2\theta]$$

$$= \frac{1}{2}\left(\frac{\sigma_1}{E} - \frac{v\sigma_2}{E}\right) + \frac{1}{2}\left(\frac{\sigma_2}{E} - \frac{v\sigma_1}{E}\right)$$
$$+ \frac{1}{2}\left[\left(\frac{\sigma_1}{E} - \frac{v\sigma_2}{E}\right) - \left(\frac{\sigma_2}{E} - \frac{v\sigma_1}{E}\right)\right]\cos 2\theta$$

But from the expressions derived in Example 2.1 the principal strains are

$$\varepsilon_1 = \frac{\sigma_1}{E} - \frac{v\sigma_2}{E} \quad \text{and} \quad \varepsilon_2 = \frac{\sigma_2}{E} - \frac{v\sigma_1}{E}$$

The strain ε_θ can therefore be expressed in terms of the principal strains as

$$\varepsilon_\theta = \tfrac{1}{2}(\varepsilon_1 + \varepsilon_2) + \tfrac{1}{2}(\varepsilon_1 - \varepsilon_2)\cos 2\theta$$

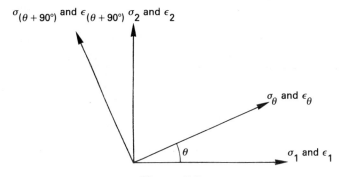

Figure 2.2

A comparison with the expression for σ_θ obtained in Chapter 1 shows that ε_θ is related to the principal strains ε_1 and ε_2 by an expression which has the same form as that relating σ_θ to the principal stresses (σ_1 and σ_2).

With the numerical data of the question $\theta = 30°$ and

$$\tfrac{1}{2}(\varepsilon_1 + \varepsilon_2) + \tfrac{1}{2}(\varepsilon_1 - \varepsilon_2)\cos 60° = 455 \times 10^{-6} \qquad\qquad (i)$$

For the perpendicular direction ($\theta = 120°$)

$$\tfrac{1}{2}(\varepsilon_1 + \varepsilon_2) - \tfrac{1}{2}(\varepsilon_1 - \varepsilon_2)\cos 60° = -32 \times 10^{-6} \qquad\qquad (ii)$$

Adding (i) and (ii) gives

$$\varepsilon_1 + \varepsilon_2 = 422 \times 10^{-6} \qquad\qquad (iii)$$

Subtracting (ii) from (i)

$$\varepsilon_1 - \varepsilon_2 = 487 \times 10^{-6} \times 2 = 974 \times 10^{-6} \qquad\qquad (iv)$$

since $\cos 60° = \tfrac{1}{2}$.

Adding (iii) and (iv)

$$\varepsilon_1 = \tfrac{1}{2}(422 + 974) \times 10^{-6} = 698 \times 10^{-6}$$

Subtracting (iv) from (iii)

$$\varepsilon_2 = \tfrac{1}{2}(422 - 974) \times 10^{-6} = -276 \times 10^{-6}$$

Using the equations given in the question

$$\sigma_1 = \frac{200 \times 10^9 \text{ N/m}^2}{1 - 0.3^2}(698 \times 10^{-6} - 0.3 \times 276 \times 10^{-6})$$

$$= 135 \text{ MN/m}^2 \text{ (tensile)} \qquad\qquad (Ans)$$

$$\sigma_2 = \frac{200 \times 10^9 \text{ N/m}^2}{1 - 0.3^2}(-276 \times 10^{-6} + 0.3 \times 698 \times 10^{-6})$$

$$= -14.7 \text{ MN/m}^2 \text{ (i.e. compressive)} \qquad\qquad (Ans)$$

2.3 Analysis of rectangular strain gauge rosette readings

A rosette of three strain gauges on the surface of a metal plate under stress gave the following strain readings: No. 1 at 0°, +0.000 592; No. 2 at 45°, +0.000 308; No. 3 at 90°, −0.000 432, the angles being measured anti-clockwise from gauge No. 1.

Determine the magnitude of the principal strains and their directions relative to the axis of gauge No. 1. If $E = 203$ GN/m² and Poisson's ratio is 1/3 find the principal stresses.

Prove any formula used. (UL)

Solution There are a number of instruments for measuring the small changes in length which occur when a metal bar or plate is stressed. Some of them depend on systems of mechanical or optical levers and

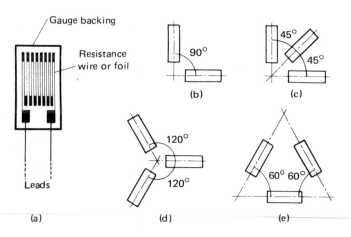

Figure 2.3

they are known as *extensometers*. Their main use is in the determination of the mechanical properties of materials and typical results are given in Volume 1, Chapter 9.

Nowadays the *electrical resistance strain gauge* is widely used to measure the strains in machine and structural elements, particularly under conditions of complex stress. The principle of this device is that the electrical resistance of a wire changes when the wire is strained. The basic gauge (Fig. 2.3(a)) consists of a grid of wire or foil mounted on a special thin backing. The gauge is usually less than 2 cm long. It is stuck to the surface of the component at the point and in the direction for which the strain is required.

In order to measure strains under conditions of complex stress two or more basic gauges are used simultaneously and they are manufactured in various configurations. Figure 2.3(b) shows two gauges mounted at right angles and this arrangement is suitable when the directions of the principal stresses are known (as in Example 2.2). Otherwise it is necessary to measure strains in at least three directions and this is done by a *rosette* of gauges mounted at convenient angles to one another. Figure 2.3(c) illustrates the configuration described in the present example, which is known as a rectangular rosette.

The mathematical relationships between the strains can be derived from the Mohr circle and Figure 2.4 gives the notation to be used. The principal strains (ε_1 and ε_2) are given by the intersections of the circle on the ε axis. Let θ be the angle between the first principal stress and the direction of the No. 1 gauge. Note that the Mohr construction uses double angles (2θ) and the radii representing the various gauge readings do not give the angles between the gauges themselves. From the diagram

$$\text{OC} = \tfrac{1}{2}(\varepsilon_0 + \varepsilon_{90}) = \tfrac{1}{2}(0.000\,592 - 0.000\,432)$$
$$= 0.000\,08$$

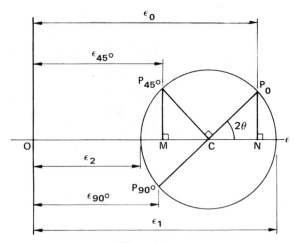

Figure 2.4

$$CN = \varepsilon_0 - OC = 0.000\,592 - 0.000\,08$$
$$= 0.000\,512$$

$$P_0N = MC = OC - \varepsilon_{45} = 0.000\,08 - 0.000\,308$$
$$= -0.000\,228$$

$$\text{Radius } P_0C = \sqrt{[CN^2 + P_0N^2]} = \sqrt{[0.000\,512^2 + (-0.000\,228)^2]}$$
$$= 0.000\,560$$

The principal strains are, therefore,

$$\varepsilon_1 = OC + \text{radius} = 0.000\,08 + 0.000\,560$$
$$= 0.000\,640$$

$$\varepsilon_2 = OC - \text{radius} = 0.000\,08 - 0.000\,560$$
$$= -0.000\,480$$

Also from the geometry of the Mohr circle the angle between the axis of gauge No. 1 and the greater principal strain is given by

$$\tan 2\theta = \frac{P_0N}{CN} = \frac{-0.000\,228}{0.000\,512} = -0.445$$

$$2\theta = -24° \quad (\text{or } 156°)$$

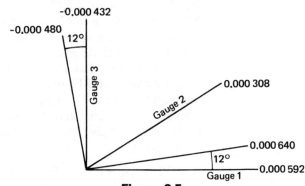

Figure 2.5

Since the No. 1 gauge strain is closer to ε_1 than ε_2 it is the (numerically) smaller angle which is required and thus

$$\theta = -12°$$

The directions of the strains are shown in Figure 2.5. The principal stresses are found as in the previous example.

$$\sigma_1 = \frac{E}{1 - v^2}(\varepsilon_1 + v\varepsilon_2)$$

$$= \frac{203 \times 10^9}{1 - (\tfrac{1}{3})^2}(0.000\,640 - \tfrac{1}{3} \times 0.000\,480)$$

$$= 109.7\,\text{MN/m}^2 \text{ (tensile)} \hspace{4cm} (Ans)$$

$$\sigma_2 = \frac{203 \times 10^9}{1 - (\tfrac{1}{3})^2}(-0.000\,480 + \tfrac{1}{3} \times 0.000\,640)$$

$$= -61.0\,\text{MN/m}^2 \text{ (i.e. compressive)} \hspace{3cm} (Ans)$$

2.4 Analysis of delta strain gauge rosette readings

> A strain gauge rosette is used for the determination of the stress condition at a point on the surface of a steel plate subjected to plane stress. The axes of the three gauges are denoted OA, OB and OC and these are 120° from one to the next. The observed strains are $\varepsilon_1 = +0.000\,554$ along OA, $\varepsilon_2 = -0.000\,456$ along OB, $\varepsilon_3 = +0.000\,064$ along OC.
>
> Determine the inclinations of the principal planes at O relative to OA and the magnitudes of the principal stresses. Determine also the strain in the direction at right angles to OA. Either graphical or analytical methods may be used. Take $E = 200\,\text{GN/}$ m^2 and Poisson's ratio $= 0.3$.
>
> Make a diagram showing the directions OA, OB and OC and the principal strains and stresses in their correct relative positions. (*UL*)

Solution An alternative to the rectangular rosette is one in which the three gauges are mounted symmetrically. This can be achieved by a 'star' configuration (Figure 2.3(d)) or by making the gauges the sides of an equilateral triangle, Figure 2.3(e), in which case it is known as a *delta rosette*.

Again the analysis can be made by reference to the Mohr circle Figure 2.6, and for consistency with earlier examples, let ε_1 and ε_2 be the principal strains, ε_0, ε_{120} and ε_{240} the given strains.

Then from the geometry of Figure 2.6, if $r = $ radius,

$$\varepsilon_0 = \text{OC} + r\cos 2\theta \hspace{5cm} \text{(i)}$$

$$\varepsilon_{120} = \text{OC} + r\cos(2\theta + 240°)$$

$$= \text{OC} + r(\cos 2\theta \cos 240° - \sin 2\theta \sin 240°)$$

$$= \text{OC} - r\cos 2\theta \cos 60° + r\sin 2\theta \sin 60° \hspace{2cm} \text{(ii)}$$

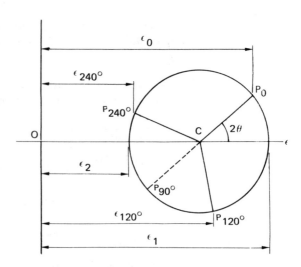

Figure 2.6

$$\varepsilon_{240} = OC + r \cos(2\theta + 480°)$$
$$= OC + r(\cos 2\theta \cos 120° - \sin 2\theta \sin 120°)$$
$$= OC - r \cos 2\theta \cos 60° - r \sin 2\theta \sin 60° \qquad \text{(iii)}$$

Adding (i), (ii) and (iii)

$$\varepsilon_0 + \varepsilon_{120} + \varepsilon_{240} = 3 \times OC + r \cos 2\theta - 2r \cos 2\theta \cos 60°$$
$$= 3 \times OC$$

(since $\cos 60° = \frac{1}{2}$).

With the values given in the question

$$OC = \tfrac{1}{3}(\varepsilon_0 + \varepsilon_{120} + \varepsilon_{240})$$
$$= \tfrac{1}{3}(0.000\,554 - 0.000\,456 + 0.000\,064)$$
$$= 0.000\,054$$

Also, subtracting (iii) from (ii)

$$\varepsilon_{120} - \varepsilon_{240} = 2r \sin 2\theta \sin 60° = \sqrt{3}r \sin 2\theta$$

(since $\sin 60° = \sqrt{3}/2$).

Substituting the numerical values,

$$r \sin 2\theta = (\varepsilon_{120} - \varepsilon_{240})/\sqrt{3} = (-0.000\,456 - 0.000\,064)/\sqrt{3}$$
$$= -0.000\,3 \qquad \text{(iv)}$$

From (i)

$$r \cos 2\theta = \varepsilon_0 - OC$$
$$= 0.000\,554 - 0.000\,054$$
$$= 0.000\,5 \qquad \text{(v)}$$

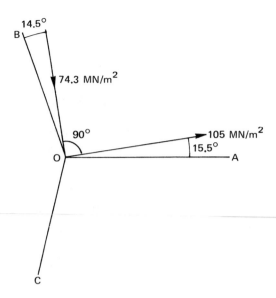

Figure 2.7

Dividing (iv) by (v) we have

$$\tan 2\theta = -0.0003/0.0005 = -0.6$$
$$2\theta = -31° \quad (\text{or } 149°)$$
$$\theta = -15.5° \qquad\qquad\qquad (Ans)$$

The principal planes are therefore 15.5° and 105.5° anti-clockwise from OA. (See Figure 2.7.)

From (i) the radius of the Mohr circle is given by

$$r = \frac{\varepsilon_0 - \text{OC}}{\cos 2\theta} = \frac{0.000\,554 - 0.000\,054}{\cos(-31°)} = 0.000\,583$$

The principal strains are therefore

$$\varepsilon_1 = \text{OC} + r = 0.000\,054 + 0.000\,583$$
$$= 0.000\,637$$
$$\varepsilon_2 = \text{OC} - r = 0.000\,054 - 0.000\,583$$
$$= -0.000\,529$$

Therefore

$$\sigma_1 = \frac{E}{1 - v^2}(\varepsilon_1 + v\varepsilon_2)$$
$$= \frac{200}{1 - 0.3^2}(0.000\,637 - 0.3 \times 0.000\,529)$$
$$= 105 \text{ MN/m}^2 \text{ (tensile)} \qquad\qquad (Ans)$$

$$\sigma_2 = \frac{E}{1-v^2}(\varepsilon_2 + v\varepsilon_1)$$

$$= \frac{200}{1-0.3^2}(-0.000\,529 + 0.3 \times 0.000\,637)$$

$$= -74.3\,\text{MN/m}^2 \text{ (i.e. compressive)} \qquad (Ans)$$

If θ is now measured from the direction of ε_1 its value for the direction at right angles to OA is $(90 - 15.5)° = 74.5°$. The strain in this direction is

$$\varepsilon = \tfrac{1}{2}(\varepsilon_1 + \varepsilon_2) + \tfrac{1}{2}(\varepsilon_1 - \varepsilon_2)\cos 2\theta$$

$$= \tfrac{1}{2}(0.000\,637 - 0.000\,529) + \tfrac{1}{2}(0.000\,637 + 0.000\,529)\cos 149°$$

$$= -0.000\,446 \qquad (Ans)$$

Alternatively, from Figure 2.6, by symmetry,

$$\text{OC} - \varepsilon_{90} = \varepsilon_0 - \text{OC} = 0.000\,5 \quad \text{(from (iv))}$$

Hence

$$\varepsilon_{90} = \text{OC} - 0.000\,5 = 0.000\,054 - 0.000\,5$$

$$= -0.000\,446 \qquad (Ans)$$

The alternative graphical methods referred to in the question are explained in the next two examples.

2.5 Graphical solution of a rectangular strain gauge rosette

Derive expressions for the principal strains in terms of ε_0, ε_{45} and ε_{90}, the readings of a rectangular rosette of strain gauges. In a particular case $\varepsilon_0 = 1500$, $\varepsilon_{45} = 200$ and $\varepsilon_{90} = -200$, the readings being in microstrain. 1 microstrain is a strain of 10^{-6}. Calculate the corresponding principal strains.

Describe a graphical method for obtaining the results and use it to check your answers.

Solution In the solution to Example 2.3 it was shown that the position of the centre of the Mohr circle of strain, Figure 2.4, is given by

$$\text{OC} = \tfrac{1}{2}(\varepsilon_0 + \varepsilon_{90})$$

With the same notation as before, the radius of the circle is

$$P_0C = \sqrt{[CN^2 + P_0N^2]}$$

$$= \sqrt{[(\varepsilon_0 - \text{OC})^2 + (\text{OC} - \varepsilon_{45})^2]}$$

$$= \sqrt{[(\varepsilon_0 - \tfrac{1}{2}\varepsilon_0 - \tfrac{1}{2}\varepsilon_{90})^2 + (\tfrac{1}{2}\varepsilon_0 + \tfrac{1}{2}\varepsilon_{90} - \varepsilon_{45})^2]}$$

$$= \sqrt{[\tfrac{1}{4}(\varepsilon_0 - \varepsilon_{90})^2 + \tfrac{1}{4}(\varepsilon_0 + \varepsilon_{90} - 2\varepsilon_{45})^2]}$$

The required expressions for the principal strains are therefore:

$$\varepsilon_1 = OC + P_0C = \tfrac{1}{2}(\varepsilon_0 + \varepsilon_{90}) + \tfrac{1}{2}\surd[(\varepsilon_0 - \varepsilon_{90})^2 + (\varepsilon_0 + \varepsilon_{90} - 2\varepsilon_{45})^2]$$

$$\varepsilon_2 = OC - P_0C = \tfrac{1}{2}(\varepsilon_0 + \varepsilon_{90}) - \tfrac{1}{2}\surd[(\varepsilon_0 - \varepsilon_{90})^2 + (\varepsilon_0 + \varepsilon_{90} - 2\varepsilon_{45})^2]$$

With the numerical data of the question

$$\varepsilon_1 = \tfrac{1}{2}(1500 - 200) + \tfrac{1}{2}\surd[(1500 + 200)^2 + (1500 - 200 - 400)^2]$$

$$= 650 + 962 = 1612 \text{ microstrain} \hspace{3cm} (Ans)$$

and

$$\varepsilon_2 = 650 - 962 = -312 \text{ microstrain} \hspace{3cm} (Ans)$$

The following graphical method of solution is based upon the principle that a circle—in this case Mohr's circle—is fully defined if three points on its circumference are given. The three points are located by the magnitudes of the strains recorded by the rosette's gauges together with the angles between the gauges.

Consider the Mohr circle of Figure 2.4 and suppose that vertical lines are drawn through the points P_0, P_{45} and P_{90}. Then, as shown in Figure 2.8 the distances of these lines from the origin O represent ε_0, ε_{45} and ε_{90}.

Taking two points at a time as shown in Figures 2.9(a) and (b) it can be seen that the arc $P_{45}P_0$ subtends an angle of 90° at the centre and 45° at the point A of the circumference, while $P_{45}P_{90}$ also subtends an angle of 45° at A.

If the construction is started at A on the vertical line through P_{45}, the points P_0 and P_{90} are located as the intersections between their vertical lines and construction lines drawn through A at angles of 45° clockwise and anti-clockwise from the vertical.

Figure 2.10 illustrates the construction at (a) with the corresponding gauge positions shown at (b).

The steps are as follows:

1. Draw, to a suitable scale, the four vertical lines representing zero strain, ε_0, ε_{45} and ε_{90}.

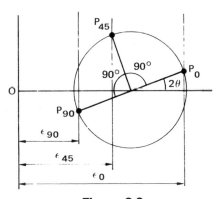

Figure 2.8

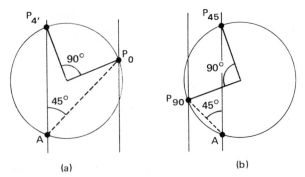

Figure 2.9

2. Mark the point A at an arbitrarily chosen position on the ε_{45} line.

3. Draw lines from A at 45° clockwise and 45° anti-clockwise to intersect the ε_0 and ε_{90} lines at P_0 and P_{90} respectively.

4. Draw the perpendicular bisectors of AP_0 and AP_{90} to intersect at C, the centre of the circle.

5. Construct the circle, centre C, passing through A, P_0 and P_{90} to intersect the ε_{45} line at P_{45}.

6. Draw a horizontal line through C and determine the principal strains by measuring the distances from the zero strain line of the points where it intersects the circle.

With appropriate changes in angles the same steps can be used to analyse the readings from a delta rosette.

Figure 2.11 shows the construction and resulting strain circle to a scale of 25 microstrain per mm for the data given in the question. The distances from O of the ends of the horizontal diameter are found to be 64.5 mm and −12.5 mm.

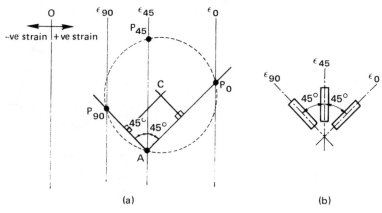

Figure 2.10

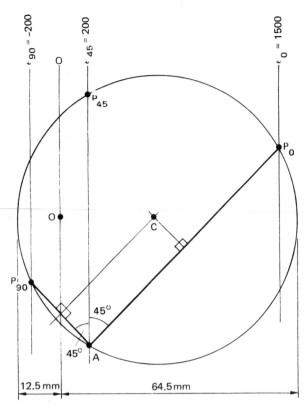

$\varepsilon_{90} = -200$ $\varepsilon_{45} = 200$ $\varepsilon_0 = 1500$

O

P_{45}

P_0

O

C

P'_{90}

45°

45°

45° A

12.5 mm   64.5 mm

Scale = 25 microstrain per mm

Figure 2.11

$\varepsilon_1 = 64.5 \times 25 = 1613$ microstrain   (*Ans*)

$\varepsilon_2 = -12.5 \times 25 = -313$ microstrain   (*Ans*)

2.6 Relationship between Mohr circles for stress and strain

Calculate the principal stresses for the state of stress given in the previous example. Show also that the Mohr circle of stress can be developed from the strain circle and hence determine the values of the principal stresses graphically.
Take $E = 80\,\text{GN/m}^2$ and $v = 0.3$.

Solution Using the answers from the last example and the relationships derived in Example 2.2 we have

$$\sigma_1 = \frac{E}{1 - v^2}(\varepsilon_1 + v\varepsilon_2)$$

$$= \frac{80 \times 10^9}{1 - 0.3^2}(1612 \times 10^{-6} - 0.3 \times 312 \times 10^{-6})$$

$$= 133.5\,\text{MN/m}^2 \quad \text{(tensile)} \qquad (Ans)$$

$$\sigma_2 = \frac{E}{1 - v^2}(\varepsilon_2 + v\varepsilon_1)$$

$$= \frac{80 \times 10^9}{1 - 0.3^2}(-312 \times 10^{-6} + 0.3 \times 1612 \times 10^{-6})$$

$$= 15.1 \text{ MN/m}^2 \quad \text{(tensile)} \hspace{3cm} (Ans)$$

Note The expressions for ε_1 and ε_2 obtained in the last solution can be substituted in the above relationships to give formulae for σ_1 and σ_2 in terms of the rosette gauge readings. The results are given in Problem 16 of this chapter and are convenient for use with programmable calculators.

It is possible to combine Mohr's circles for strain and stress in a single diagram by a suitable adjustment of scales. Consider a point in a material where the principal strains are ε_1, ε_2 and the corresponding principal stresses are σ_1, σ_2.

The general state of strain at the point can be represented by the Mohr's strain circle, Figure 2.12(a) and the state of stress by the Mohr's stress circle, Figure 2.12(b).

In these diagrams the lengths OC and OC_1 represent $\frac{1}{2}(\varepsilon_1 + \varepsilon_2)$ and $\frac{1}{2}(\sigma_1 + \sigma_2)$ respectively, i.e.

$$OC \times \text{strain scale} = \tfrac{1}{2}(\varepsilon_1 + \varepsilon_2)$$

and

$$OC_1 \times \text{stress scale} = \tfrac{1}{2}(\sigma_1 + \sigma_2)$$

It is convenient to draw the circles concentrically and for this OC and OC_1 must be the same length. Hence:

$$\frac{\text{stress scale}}{\text{strain scale}} = \frac{\sigma_1 + \sigma_2}{\varepsilon_1 + \varepsilon_2}$$

But, with the relationships quoted at the beginning of this solution,

$$\sigma_1 + \sigma_2 = \frac{E}{1 - v^2}(\varepsilon_1 + v\varepsilon_2) + \frac{E}{1 - v^2}(\varepsilon_2 + v\varepsilon_1)$$

$$= \frac{E}{1 - v^2}(\varepsilon_1 + \varepsilon_2)(1 + v)$$

$$= \frac{E}{1 - v}(\varepsilon_1 + \varepsilon_2)$$

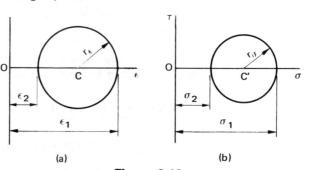

(a)   (b)

Figure 2.12

Therefore, to achieve concentricity, it is required that

$$\text{stress scale} = \frac{E}{1-v} \times \text{strain scale}$$

A consequence of adopting this ratio of scales for the two circles is that their radii (r_ε and r_σ) will have a specific ratio. From Figures 2.12(a) and (b),

$$r_\varepsilon \times \text{strain scale} = \tfrac{1}{2}(\varepsilon_1 - \varepsilon_2)$$

and

$$r_\sigma \times \text{stress scale} = \tfrac{1}{2}(\sigma_1 - \sigma_2)$$

Therefore by division

$$\frac{r_\sigma}{r_\varepsilon} = \frac{\sigma_1 - \sigma_2}{\varepsilon_1 - \varepsilon_2} \times \frac{\text{strain scale}}{\text{stress scale}}$$

But

$$\sigma_1 - \sigma_2 = \frac{E}{1-v^2}(\varepsilon_1 + v\varepsilon_2) - \frac{E}{1-v^2}(\varepsilon_2 + v\varepsilon_1)$$

$$= \frac{E}{1-v^2}(\varepsilon_1 - \varepsilon_2)(1-v)$$

$$= \frac{E}{1+v}(\varepsilon_1 - \varepsilon_2)$$

Furthermore,

$$\frac{\text{strain scale}}{\text{stress scale}} = \frac{1-v}{E}$$

Using these two results the ratio of the radii becomes

$$\frac{r_\sigma}{r_\varepsilon} = \frac{E}{1+v} \times \frac{1-v}{E}$$

i.e. radius of stress circle $= \dfrac{1-v}{1+v} \times$ radius of strain circle

The numerical data can now be used to construct the stress circle corresponding to the strain circle of Figure 2.11. The scale of that diagram was 25 microstrain/mm and hence:

$$\text{stress scale} = \frac{E}{1-v} \times \text{strain scale}$$

$$= \frac{80 \times 10^9 \,\text{N/m}^2}{1-0.3} \times 25 \times 10^{-6} \,\text{per mm}$$

$$= 2.86 \,\text{MN/m}^2 \,\text{per mm}$$

By measurement of Figure 2.11 the radius of the strain circle is

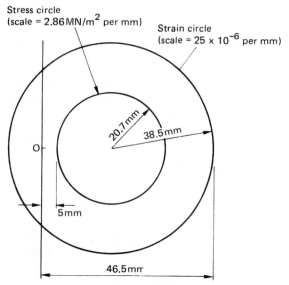

Stress circle
(scale = 2.86 MN/m^2 per mm)

Strain circle
(scale = 25 x 10^{-6} per mm)

20.7 mm

38.5 mm

O

5 mm

46.5 mm

Figure 2.13

38.5 mm and hence:

$$\text{radius of stress circle} = \frac{1-v}{1+v} \times \text{radius of strain circle}$$

$$= \frac{1-0.3}{1+0.3} \times 38.5 \text{ mm}$$

$$= 20.7 \text{ mm}$$

The two circles, with their separate scales, are shown in Figure 2.13. The measured distances from the vertical axis of the ends of the horizontal diameter of the stress circle are 46.5 mm and 5 mm. Therefore

$$\sigma_1 = 46.5 \text{ mm} \times 2.86 \text{ MN/m}^2 \text{ per mm}$$

$$= 133.0 \text{ MN/m}^2 \text{ (tensile)} \tag{Ans}$$

and

$$\sigma_2 = 5 \text{ mm} \times 2.86 \text{ MN/m}^2 \text{ per mm}$$

$$= 14.3 \text{ MN/m}^2 \text{ (tensile)} \tag{Ans}$$

These results confirm the answers obtained at the beginning of this solution within the limits of drawing accuracy.

2.7 Change in volume of rectangular bar

A cylindrical bar 150 mm long by 75 mm diameter is subjected to an axial compressive load of 450 kN. Taking $E = 200$ GN/m^2 and $v = 0.3$ calculate the change in volume of the bar. If all lateral swelling of the bar is prevented what is the pressure on the side of the bar and the change in volume?

Solution If a cylindrical bar, length l and diameter d, has these dimensions increased to $(l + \delta l)$ and $(d + \delta d)$ respectively, the change in volume is

$$\delta V = \tfrac{1}{4}\pi(d + \delta d)^2(l + \delta l) - \tfrac{1}{4}\pi d^2 l$$
$$= \tfrac{1}{4}\pi(d^2 l + 2dl\delta d + l\delta d^2 + d^2\delta l + 2d\delta d\delta l + \delta l\delta d^2 - d^2 l)$$
$$= \tfrac{1}{4}\pi(2dl\delta d + d^2\delta l)$$

(neglecting, as usual, the products of small quantities).
Hence

$$\delta V = \tfrac{1}{4}\pi d^2 l\left(2\frac{\delta d}{d} + \frac{\delta l}{l}\right)$$

$$\frac{\delta V}{V} = (\text{twice the diametral strain} + \text{the longitudinal strain})$$

$$\text{Longitudinal stress} = \frac{450 \times 10^3 \text{ N}}{\tfrac{1}{4}\pi \times (75 \times 10^{-3} \text{ m})^2}$$
$$= 102 \text{ MN/m}^2 \text{ (compressive)}$$

$$\text{Longitudinal strain} = \text{stress}/E = -\frac{102 \times 10^6 \text{ N/m}^2}{200 \times 10^9 \text{ N/m}^2}$$
$$= -0.000\,510$$

The diametral (lateral) strain $= -\nu \times \text{longitudinal strain}$
$$= -0.3(-0.000\,510)$$
$$= 0.000\,153$$

Thus, the change in volume is

$$\delta V = V \text{ (twice the diametral strain} + \text{the longitudinal strain)}$$
$$= \left[\frac{\pi}{4}(75 \text{ mm})^2(150 \text{ mm})\right][2(0.000\,153) + (-0.000\,510)]$$
$$= 135 \text{ mm}^3$$

Let p MN/m^2 be the pressure on the surface of the bar necessary to prevent all lateral swelling. By symmetry, it is the same at all points on the surface. Hence, both lateral principal stresses are equal to $-p$, and if σ_1 is the longitudinal stress then

$$\sigma_1 = -102 \qquad \sigma_2 = \sigma_3 = -p$$

Since the strain is zero in the direction of any diameter,

$$\frac{\sigma_2}{E} - \frac{\nu\sigma_3}{E} - \frac{\nu\sigma_1}{E} = 0$$

or

$$\frac{(-p)}{E} - \frac{\nu(-p)}{E} = \frac{\nu\sigma_1}{E}$$
$$p(\nu - 1) = \nu\sigma_1 \quad \text{or} \quad p = \nu\sigma_1/(\nu - 1)$$

The required pressure is therefore

$$\left(\frac{v}{v-1}\right) \times \sigma_1 = \frac{0.3 \times (-102)}{(0.3-1)} = 43.7 \, \text{MN/m}^2 \qquad (Ans)$$

since the diametral strain is zero, the change in volume is

$$\delta V = V \times \text{longitudinal strain}$$

$$= V \times \left(\frac{\sigma_1}{E} - \frac{v\sigma_2}{E} - \frac{v\sigma_3}{E}\right)$$

$$= \frac{V}{E} \times (-102 + 2vp)$$

$$= \left[\frac{\frac{1}{4}\pi \times (75 \, \text{mm})^2 \times 150 \, \text{mm}}{200 \, \text{GN/m}^2}\right][-102 + (2 \times 0.3 \times 43.7)] \, \text{MN/m}^2$$

$$= -251 \, \text{mm}^3 \qquad (Ans)$$

i.e. a decrease of 251 mm³.

<table>
<tr><td>

2.8 Bulk modulus and its relationship to the modulus of elasticity

</td><td>

Define bulk modulus and derive an expression relating it to Young's modulus.

Calculate the change in volume of a 15 cm cube of steel when it is immersed to a depth of 300 m in sea-water which weighs 10 kN/m³

$$E = 200 \, \text{GN/m}^2 \qquad v = 0.29$$

</td></tr>
</table>

Solution   When a body is subjected to the same pressure at all points on its surface the volumetric strain is proportional to this stress (or pressure). The *bulk modulus* is defined as

$$\frac{\text{stress}}{\text{volumetric strain}}$$

and is usually denoted by K.

Since the volumetric strain can also be obtained as the sum of the three principal strains (which involve E and v) then a relationship between K, E and v can be deduced.

Suppose a rectangular block is subjected to a pressure p on all faces. The principal stresses are

$$\sigma_1 = \sigma_2 = \sigma_3 = -p$$

and each principal strain is

$$\frac{\sigma_1}{E} - \frac{v\sigma_2}{E} - \frac{v\sigma_3}{E} = -\left(\frac{p}{E} - \frac{2vp}{E}\right)$$

The volumetric strain is thus

$$-3 \times \left(\frac{p}{E} - \frac{2vp}{E} \right) = -\frac{3p}{E}(1 - 2v) \qquad (i)$$

By definition, however,

$$\text{volumetric strain} = \frac{\text{stress}}{K} = \frac{-p}{K} \qquad (ii)$$

Equating (i) and (ii),

$$-\frac{p}{K} = -\frac{3p}{E}(1 - 2v)$$

$$E = 3K(1 - 2v) \qquad (Ans)$$

With the values given in the question,

$$K = \frac{E}{3(1 - 2v)} = \frac{200 \times 10^9 \, \text{N/m}^2}{3[1 - (2 \times 0.29)]}$$

$$= 159 \, \text{GN/m}^2$$

The pressure at the given depth is

$$p = 300 \, \text{m} \times 10 \, \text{kN/m}^3 = 3 \, \text{MN/m}^2$$

The volumetric strain $\delta V / V = \text{stress}/K$

$$= \frac{-3 \times 10^6 \, \text{N/m}^2}{159 \times 10^9 \, \text{N/m}^2}$$

$$= -1.89 \times 10^{-5}$$

The change in volume $\delta V = V \times 1.89 \times 10^{-5}$

$$= (15 \times 10^{-2} \, \text{m})^3 \times (-1.89 \times 10^{-5})$$

$$= -63.8 \, \text{mm}^3 \qquad (Ans)$$

2.9 Compressibility of a liquid and the change in capacity of a thin-walled tube

A copper tube 50 mm internal diameter, 1 m long and 1 mm thick, has closed ends and is filled with water under pressure. Neglecting any distortion of the end plates, determine the alteration in pressure when an additional 3 cm³ of water is pumped into the tube.

Modulus of elasticity for copper = 100 GN/m²
Poisson's ratio = 0.3
Bulk modulus for water = 2 GN/m² (UL)

Solution It was shown in Volume 1, Chapter 10 that, with certain assumptions, the proportional increase in the capacity of a thin-walled cylinder due to an internal pressure P is

$$\frac{Pd}{4tE}(5 - 4v)$$

where d is the cylinder diameter and t the wall thickness.

With the numerical values given in the question,

$$\frac{Pd}{4tE}(5 - 4v) = \frac{P \text{ N/m}^2 \times 0.05 \text{ m}}{4 \times 0.001 \text{ m} \times (100 \times 10^9 \text{ N/m}^2)}(5 - 1.2)$$

$$= 4.75 \times 10^{-10} \times P \tag{i}$$

The proportional decrease in the volume of water originally in the cylinder is

$$\text{Volumetric strain} = \frac{\text{pressure}}{\text{bulk modulus}} = \frac{P \text{ N/m}^2}{2 \times 10^9 \text{ N/m}^2}$$

$$= 5 \times 10^{-10} \times P \tag{ii}$$

The sum of expressions (i) and (ii) equals the ratio of the additional volume of water to the original capacity of the cylinder. Hence

$$(4.75 \times 10^{-10} \times P) + (5 \times 10^{-10} \times P) = \frac{3 \times 10^{-6} \text{ m}^3}{\frac{1}{4}\pi \times (0.05 \text{ m})^2 \times (1 \text{ m})}$$

$$= 1.53 \times 10^{-3}$$

and, multiplying through by 10^{10},

$$9.75P = 1.53 \times 10^7 \qquad P = 1.57 \times 10^6 \text{ N/m}^2$$

The alteration in pressure is 1.57 MN/m^2 (or 15.7 bar) $\qquad$ (*Ans*)

2.10 Relationship between the modulus of rigidity, the modulus of elasticity and Poisson's ratio

Prove the relationship between Young's modulus, the modulus of rigidity and Poisson's ratio.

When a bar of certain material, 50 mm diameter, is subjected to an axial pull of 250 kN the extension on a gauge length of 200 mm is 338 μm and the decrease in diameter is 26 μm. Calculate Young's modulus of elasticity, Poisson's ratio, the shear modulus and bulk modulus for this material.

Solution Consider a cube ABCD which is distorted by (pure) shear stress τ to a position AB'C'D as shown in Figure 2.14. For pure shear the principal stresses are $\sigma_1 = +\tau$ and $\sigma_2 = -\tau$ acting along the diagonals AC and BD respectively.

Since the strain along the diagonal AC can be expressed in terms of the shear modulus or, alternatively, in terms of modulus of elasticity and Poisson's ratio, then a relationship between these three quantities can be deduced.

If CE is a perpendicular from C on to the new position of the

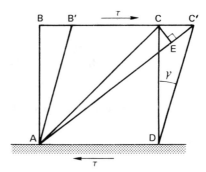

Figure 2.14

diagonal AC' then, since the distortion is small, the strain along AC' is

$$\frac{EC'}{AC} = \frac{CC' \cos EC'C}{AC} = \frac{CC' \cos 45°}{AC} \text{ (approximately)}$$

$$= \frac{CC' \cos 45°}{(DC/\cos 45°)} \quad \text{(since } DC/AC = \cos 45°)$$

$$= \frac{CC'}{DC} \cos^2 45°$$

$$= \frac{1}{2}\left(\frac{CC'}{DC}\right)$$

However CC'/DC is the shear strain γ where

$$\gamma = \text{shear stress}/G = \tau/G$$

Hence, strain along AC' is $\tau/2G$. \hfill (i)

In terms of the principal stresses $\sigma_1(=\tau)$ and $\sigma_2(=-\tau)$,

$$\text{diagonal strain} = \frac{\sigma_1}{E} - \frac{v\sigma_2}{E} = \frac{\tau}{E} - \frac{v(-\tau)}{E}$$

$$= \frac{\tau}{E}(1+v) \hfill \text{(ii)}$$

Since (i) and (ii) represent the same strain,

$$\frac{\tau}{2G} = \frac{\tau}{E}(1+v)$$

$$E = 2G(1+v)$$

Combining this with the result obtained in Example 2.8,

$$E = 2G(1+v) = 3K(1-2v)$$

and, if any two of the four elastic constants E, G, K, v are given, the others can be calculated.

From the data given in the question,

$$E = \frac{\text{stress}}{\text{strain}} = \frac{250 \times 10^3\,\text{N}}{\frac{1}{4}\pi \times (50 \times 10^{-3}\,\text{m})^2} \times \frac{200 \times 10^{-3}\,\text{m}}{338 \times 10^{-6}\,\text{m}}$$

$$= 75.4\,\text{GN/m}^2 \qquad\qquad (Ans)$$

Poisson's ratio $v = $ the absolute value of lateral
strain/longitudinal strain

$$= \frac{26 \times 10^{-6}\,\text{m}}{50 \times 10^{-3}\,\text{m}} \times \frac{200 \times 10^{-3}\,\text{m}}{338 \times 10^{-6}\,\text{m}}$$

$$= 0.3078 \qquad\qquad (Ans)$$

Shear modulus $G = E/2(1 + v)$

$$= \frac{75.4 \times 10^9\,\text{N/m}^2}{2(1 + 0.3078)} = 28.8\,\text{GN/m}^2 \qquad\qquad (Ans)$$

Bulk modulus $K = E/3(1 - 2v)$

$$= \frac{75.4 \times 10^9\,\text{N/m}^2}{3[1 - (2 \times 0.3078)]} = 65.4\,\text{GN/m}^2 \qquad\qquad (Ans)$$

2.11 Deflection and twist of a round bar under bending and torsion

A hollow round bar deflects 6 mm under a central load of 10 kN when it is used as a simply-supported beam over a span of 3.6 m. Calculate the angle of twist in this length when the bar is subjected to a torque of 5 kN m. Poisson's ratio $= \frac{3}{11}$.

Solution Since $E = 2G(1 + v)$

$$G = \frac{E}{2(1 + v)} = \frac{E}{2(1 + \frac{3}{11})} = \frac{11E}{28}$$

Also, for all circular sections, $J = 2I$.

In the 'beam' case the central deflection Δ is $WL^3/48EI$ and, using the given numerical values,

$$EI = \frac{WL^3}{48\Delta} = \frac{(10 \times 10^3\,\text{N}) \times (3.6\,\text{m})^3}{48 \times (6 \times 10^{-3}\,\text{m})} = 1.62\,\text{MN m}^2$$

Hence

$$GJ = \frac{11E}{28} \times 2I = \frac{11}{14}EI = \frac{11 \times 1.62}{14} = 1.27\,\text{MN m}^2$$

From the torsion equation,

$$\theta = \frac{TL}{GJ} = \frac{(5 \times 10^3\,\text{N m}) \times (3.6\,\text{m})}{1.27\,\text{MN m}^2}$$

$$= 0.014\,14\,\text{rad or } 0.81° \qquad\qquad (Ans)$$

Problems

1 What is meant by the term 'principal stress'?

A tube, 20 mm internal diameter and wall thickness 5 mm, is subjected to an axial tensile load of 30 kN and a twisting moment of 750 N m. Calculate the maximum principal stress in the tube. What stress acting alone would produce a strain equal to the maximum principal strain? Take Poisson's ratio $v = 0.3$.

Answer Maximum principal stress = 218 MN/m²; the maximum principal strain is equal to that caused by a simple tension of 261 MN/m².

2 Working from first principles find

(a) the normal stress,
(b) the tangential stress,
(c) the resultant stress, on the interface AB of the mild steel plate stressed as shown in Figure 2.15.

If the dimensions of the plate were 250 mm by 125 mm before straining, estimate the change in area of the plate.

$$E = 206 \text{ GN/m}^2; \quad v = 0.28 \qquad\qquad (IMechE)$$

Answer (a) 45 MN/m² (tensile); (b) 52.0 MN/m² (compressive); (c) 68.7 MN/m²; change in area, 3.28 mm² (increase).

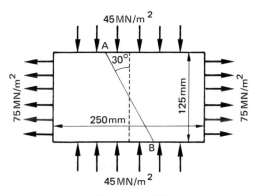

Figure 2.15

3 The principal strains at a point on the surface of a loaded elastic member are $+2 \times 10^{-4}$ and -10×10^{-4}. What are the principal stresses at the same point? Poisson's ratio $v = 0.3$. Young's modulus of elasticity $E = 200$ GN/m² $\qquad (PCL)$
Answer 22.0 and 207 MN/m² (both compressive).

4 The strains measured by electric resistance strain gauges in the 120° rosette shown in Figure 2.16 are as follows:

$$\varepsilon_1 = -1.0 \times 10^{-4}; \qquad \varepsilon_2 = +2.0 \times 10^{-4}; \qquad \varepsilon_3 = +8.0 \times 10^{-4}$$

Find the magnitude and direction of the principal strains, and the

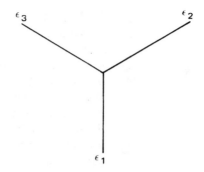

Figure 2.16

magnitude of the maximum shear strain. (*PCL*)
Answer 8.29×10^{-4} and -2.29×10^{-4} at angles of 69.6° and 159.6° respectively anti-clockwise from ε_1; 10.58×10^{-4}.

5 Find the values of v if (a) $E = 2\frac{1}{2}G$; (b) $E = K$; (c) $E = 1\frac{1}{2}K$; (d) $K = 2G$.
Answer (a) 0.25; (b) 0.333; (c) 0.25; (d) 0.286.

6 Establish the following results, assuming the relationships between the elastic moduli given at the head of this chapter.

(a) $G = \dfrac{3K}{2}\left(\dfrac{1 - 2v}{1 + v}\right)$

(b) $v = \dfrac{E - 2G}{2G} = \dfrac{3K - E}{6K} = \dfrac{3K - 2G}{6K + 2G}$

(c) $E = 9GK/(3K + G)$

(d) $G = 3KE/(9K - E)$

(e) $K = GE/(9G - 3E)$

7 A bar of magnesium alloy, 32 mm diameter, is tested on a gauge length of 200 mm in tension and also in torsion. A tensile load of 50 kN produces an extension of 0.27 mm and a torque of 200 N m produces a twist of 1.22°. Calculate Young's modulus, the shear modulus, the bulk modulus and Poisson's ratio for this material.
Answer $E = 46.0 \text{ GN/m}^2$; $G = 18.3 \text{ GN/m}^2$; $K = 32.0 \text{ GN/m}^2$; $v = 0.26$.

8 A circle 30 cm in diameter is scribed on a mild steel plate before it is stressed as shown in Figure 2.17. After stressing, the circle deforms to an ellipse. Calculate the lengths of the major and minor axes of the ellipse and also find their directions.
$E = 206 \text{ GN/m}^2$; $v = 0.28$ (*IMechE*)

Answer Major axis, 30.0109 cm; minor axis, 29.9979 cm; these axes are 18.3° and 108.3° from the vertical centre-line (measured in a clockwise sense).

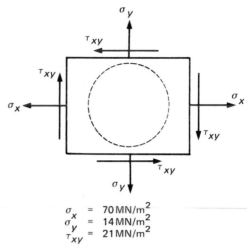

$$\sigma_x = 70\,\text{MN/m}^2$$
$$\sigma_y = 14\,\text{MN/m}^2$$
$$\tau_{xy} = 21\,\text{MN/m}^2$$

Figure 2.17

9 Derive an expression showing the relationship between the following elastic constants: Young's modulus, modulus of rigidity and Poisson's ratio.

A vertical rod of length L and diameter d is fixed at its upper end and carries a load W at the lower end. The extension of the rod due to W is x. If a torque T, applied in a horizontal plane at the lower end of the rod, gives an angle of twist θ radians show that Poisson's ratio may be obtained from the formula

$$v = \frac{Wd^2\theta}{16Tx} - 1 \qquad\qquad (IMechE)$$

10 Deduce the relation between the modulus of elasticity, the modulus of rigidity and Poisson's ratio.

A hollow shaft of 75 mm external diameter and 50 mm internal diameter twists through an angle of 0.6° in a length of 1.2 m when subjected to an axial twisting moment of 1 kN m. Estimate the deflection at the centre of the shaft due to its own weight when placed in a horizontal position on supports 1.2 m apart.

Weight of shaft 190 N per metre length. Poisson's ratio 0.3.
Answer 0.0344 mm. (UL)

11 For an elastic material, express Poisson's ratio in terms of the bulk modulus and the modulus of rigidity, and prove the derivation of the expression.

Determine the percentage change in volume of a steel bar 8 cm square in section and 1 m long when subjected to an axial compressive load of 20 kN. What change in volume would a 10 cm cube of steel suffer at a depth of 5 km in sea-water?

For steel, modulus of elasticity is 200 GN/m^2 and modulus of rigidity is 80 GN/m^2. (UL)
Answer Poisson's ratio $v = (3K - 2G)/(6K + 2G)$; volume of

bar decreases by 0.000 781 per cent; cube decreases by 377 mm³
(assuming sea-water weighs 10.06 kN/m³).

12 A brass plate 3 mm thick is stretched in the plane of the
plate in two directions at right angles. An extensometer arranged
in the x-direction gave an extension of 36×10^{-3} mm on a 50 mm
gauge length, and another extensometer, in the y-direction, gave
an extension of 17×10^{-3} mm on a 100 mm gauge length. If the
normal stress on a plane making an angle θ with the y-axis is
59.33 MN/m², find θ. Also find the decrease in the thickness of
the plate. Take E for brass as 80 GN/m² and $v = 0.3$.

(*IMechE*)

Answer   30°; 1.144×10^{-3} mm.

13 An element of elastic material is acted upon by three
principal stresses and the three principal strains ε_x, ε_y and ε_z are
measured. Show that the principal stress in the x-direction is
given by

$$\sigma_x = \alpha\Delta + 2G\varepsilon_x$$

where

$$\alpha = vE/(1 + v)(1 - 2v)$$

Δ is the volumetric strain;
G is the modulus of rigidity and v is Poisson's ratio.
In a certain test the principal strains were found to be

$$+0.000\,71; \qquad +0.001\,40; \qquad -0.001\,85$$

Determine the three principal stresses. Take $E = 200$ GN/m² and
Poisson's ratio as 0.3. (*UL*)
Answer   139 MN/m² (tensile); 245 MN/m² (tensile) and 255 MN/
m² (compressive).

14 If, under two-dimensional conditions of stress, a mean stress
σ_m and shear stress σ_s are defined as

$$\sigma_m = \tfrac{1}{2}(\sigma_1 + \sigma_2) \quad \text{and} \quad \sigma_s = \tfrac{1}{2}(\sigma_1 - \sigma_2)$$

show that σ_s and σ_m are respectively the radius of the Mohr
circle of stress and the distance of its centre from the origin.
   Show also that the principal stresses are given by $\sigma_m \pm \sigma_s$. If ε_m
and ε_s are similarly defined in terms of the principal strains ε_1
and ε_2 prove that

$$\sigma_m = \left(\frac{E}{1 - v}\right)\varepsilon_m \quad \text{and} \quad \sigma_s = \left(\frac{E}{1 + v}\right)\varepsilon_s$$

The principal strains at a point on the surface of an aircraft
structural component are measured as 0.001 15 and $-0.000\,65$.
Determine ε_m and ε_s and hence calculate the principal stresses.
$E = 91$ GN/m² and Poisson's ratio is 0.3.
Answer   0.000 25 and 0.000 90; 95.5 MN/m² (tensile) and
30.5 MN/m² (compressive).

15 In a rectangular rosette of three strain gauges the second and third gauges make angles of 45° and 90° respectively with the first. If the strains in these directions are denoted by ε_0, ε_{45} and ε_{90} show that the principal strains are given by

$$\tfrac{1}{2}(\varepsilon_0 + \varepsilon_{90}) \pm \tfrac{1}{2}\sqrt{[(\varepsilon_0 - \varepsilon_{90})^2 + (\varepsilon_0 + \varepsilon_{90} - 2\varepsilon_{45})^2]}$$

and their directions relative to that of the first gauge are given by

$$\tan 2\theta = (\varepsilon_0 + \varepsilon_{90} - 2\varepsilon_{45})/(\varepsilon_0 - \varepsilon_{90})$$

In a particular case $\varepsilon_0 = 0.001\,00$, $\varepsilon_{45} = 0.000\,15$ and $\varepsilon_{90} = -0.000\,20$. Determine the principal strains and principal stresses taking $E = 200\ \text{GN/m}^2$ and Poisson's ratio $= \tfrac{1}{3}$.

Make a diagram showing the directions of the principal stresses relative to the axes of the three gauges.
Answer +0.001 05 and −0.000 25; 217.6 MN/m² (tensile) and 22.5 MN/m² (tensile).

16 Show, with the notation of the previous question, that the principal stresses can be determined from the readings of a rectangular strain gauge rosette by the formula

$$\frac{E}{2(1 - v)} (\varepsilon_0 + \varepsilon_{90}) \pm \frac{E}{2(1 + v)} \sqrt{[(\varepsilon_0 - \varepsilon_{90})^2 + (\varepsilon_0 + \varepsilon_{90} - 2\varepsilon_{45})^2]}$$

Obtain the corresponding expression for a delta rosette.
Answer

$$\frac{E}{3(1 - v)} (\varepsilon_0 + \varepsilon_{60} + \varepsilon_{120})$$

$$\pm \frac{E}{3(1 + v)} \sqrt{[(2\varepsilon_0 - \varepsilon_{60} - \varepsilon_{120})^2 + 3(\varepsilon_{120} - \varepsilon_{60})^2]}$$

17 A solid circular shaft 50 mm diameter is subject to an axial force P and a torque T. On its surface is a delta rosette of strain gauges which reads, in microstrain

$$\varepsilon_0 = +750; \qquad \varepsilon_{60} = -414; \qquad \varepsilon_{120} = +452$$

The zero angle gauge is in the direction of the axis of the shaft. By drawing a Mohr strain circle, or otherwise, determine the values of P and T.

Take $E = 200\ \text{GN/m}^2$, $v = 0.3$.
Answer 295 kN; 1887 N m.

18 Three gauges of a strain gauge rosette are arranged in a delta formation, having angles of 0°, 60° and 120° respectively to an arbitrary axis. The strains measured by the three gauges at one instant are

$$\varepsilon_0 = 300 \times 10^{-6}; \qquad \varepsilon_{60} = 75 \times 10^{-6}; \qquad \varepsilon_{120} = 75 \times 10^{-6}$$

Determine the state of stress where the rosette gauge is attached. Take $E = 200\ \text{GN/m}^2$, $v = 0.3$.

Answer Principal strains are 300×10^6 along, and zero at right angles to, the arbitrary axis. Principal stresses are $65.9\,\text{MN/m}^2$ (tensile) and $19.8\,\text{MN/m}^2$ (tensile) respectively.

19 A thin-walled cylinder of length l and internal diameter d is made of plates of thickness t and subjected to an internal pressure of intensity p. Show that, if the effect of the ends is neglected, the increase in volume due to the pressure is

$$\pi p d^3 l (5 - 4v)/16tE$$

where E is the modulus of elasticity and v is Poisson's ratio.

A steel cylinder 0.9 m long, 150 mm internal diameter, plate thickness 5 mm is subjected to an internal pressure of $7\,\text{MN/m}^2$; the increase in volume due to the pressure is $15.45\,\text{cm}^3$. Find the values of Poisson's ratio and the modulus of rigidity. Assume $E = 206\,\text{GN/m}^2$. *(UL)*
Answer $v = 0.292$; $G = 80\,\text{GN/m}^2$.

3

Strain energy and theories of failure

Resilience

When a force (or couple) is applied to an elastic material it causes a linear (or angular) displacement and work is done. This work is stored in the material and reappears when the force (or couple) is removed. It is known as *strain energy* or *resilience* and is denoted by U.

Uniform stress

The strain energy stored per unit volume of material when it is subjected to a uniform direct stress σ is $\sigma^2/2E$. If σ is the proof stress this result is called the *modulus of resilience*. For a uniform shear stress τ the corresponding result is $\tau^2/2G$.

Bending

For a beam (or cantilever) length L, in which M and EI are the bending moment and flexural rigidity respectively at distance x from one end

$$U = \int_0^L \frac{M^2}{2EI}\,dx$$

Torsion

For a circular shaft, length L, subjected to a gradually applied torque T, the strain energy is

$$U = \frac{T^2L}{2GJ}$$

If the shaft is hollow, with external and internal diameters D and d respectively, and the maximum shear stress is τ, the *average* strain energy per unit volume is

$$\frac{\tau^2}{4G}\left(\frac{D^2 + d^2}{D^2}\right)$$

and for a solid shaft this becomes $\tau^2/4G$.

Impact loads

If a mass falls freely on to a bar, beam or spring, stress waves pass through the material and the determination of the maximum stress induced is complicated. Approximate results can be obtained, however, by consideration of energy changes. It is usual to assume that the decrease in potential energy of the falling mass is taken up as strain energy by the material of the bar, beam or spring and that the maximum stress is the same as that for the same strain energy under static loading.

The maximum stress σ produced by a mass of weight W which falls a distance h before straining a uniform bar, length L and cross-sectional area A, is

$$\sigma = \frac{W}{A}\left(1 + \sqrt{\left[1 + \frac{2AEh}{WL}\right]}\right)$$

If the maximum extension is negligible compared with h then

$$\sigma = \sqrt{\frac{2EWh}{AL}}$$

Equivalent static load

In problems where a mass whose weight is W falls on to a beam or cantilever it is convenient to calculate the equivalent weight W_E which would produce the same maximum stress and deflection if gradually applied at the same point. This is done by assuming that the amounts of strain energy are the same in the two cases.

In the particular case of a suddenly applied load W, $W_E = 2W$.

Theories of elastic failure

If σ_1, σ_2 and σ_3 are the principal stresses (in descending order of magnitude) in a three-dimensional stress system, and the yield stress in simple tension is σ_Y, the criterion for the onset of yielding according to the various theories is:

(a) Greatest principal stress theory:

$\sigma_1 = \sigma_Y$

(b) Greatest principal strain theory:

$\sigma_1 - v(\sigma_2 + \sigma_3) = \sigma_Y$

(c) Maximum shear stress theory:

$\sigma_1 - \sigma_3 = \sigma_Y$

(d) Total strain energy theory:

$\sigma_1^2 + \sigma_2^2 + \sigma_3^2 - 2v(\sigma_1\sigma_2 + \sigma_2\sigma_3 + \sigma_3\sigma_1) = \sigma_Y^2$

(e) Shear strain energy (Mises–Hencky) theory:

$(\sigma_1 - \sigma_2)^2 + (\sigma_2 - \sigma_3)^2 + (\sigma_3 - \sigma_1)^2 = 2\sigma_Y^2$

Worked Examples

3.1 Strain energy in tension

Derive an expression for the strain energy of a uniform bar in tension in terms of the stress σ, the modulus of elasticity E and the volume of the bar.

Calculate the strain energy in a bar 3 m long and 50 mm diameter when it is subjected to a tensile load of 200 kN. $E = 206$ GN/m^2.

Solution Suppose a uniform bar, length L and cross-sectional area A, is subjected to a gradually applied tensile load P. If the limit of proportionality is not exceeded, then the extension x is proportional to P and the graph of load against extension (Figure 3.1) is a straight line. Hence the strain energy is

$U =$ work done in stretching the bar (which is represented by the area under the graph of load against extension)

$= $ *average* force $\times$ total extension

$= \frac{1}{2}Px$

But $P = \sigma A$, and $x = L \times$ strain $= L\sigma/E$. Thus

$$U = \frac{1}{2} \times \sigma A \times \frac{L\sigma}{E}$$

$$= \frac{\sigma^2}{2E} \times AL$$

$$= \frac{\sigma^2}{2E} \times (\text{volume of the bar}) \qquad \text{(i)}$$

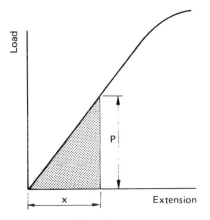

Figure 3.1

The strain energy per unit volume is

$$\frac{\sigma^2}{2E} = \tfrac{1}{2} \times \text{stress} \times \text{strain}$$

(ii)

since strain $= \sigma/E$.

Since strain energy is equal to work done, its units are newton-metres (N m). The newton-metre has the name *joule* (J).

For the bar given in the question,

$$\text{Stress } \sigma = \frac{(200 \times 10^3\,\text{N})}{\tfrac{1}{4}\pi \times (50 \times 10^{-3}\,\text{m})^2} = \frac{4 \times 200 \times 10^3}{\pi \times 2\,500 \times 10^{-6}}\,\text{N/m}^2$$

$$= 101.9\,\text{MN/m}^2$$

$$\text{Volume} = \tfrac{1}{4}\pi \times (50 \times 10^{-3}\,\text{m})^2 \times 3\,\text{m} = 5.89 \times 10^{-3}\,\text{m}^3$$

The strain energy is, therefore,

$$U = \frac{\sigma^2}{2E} \times \text{volume} = \frac{(101.9 \times 10^6\,\text{N/m}^2)^2 \times (5.89 \times 10^{-3}\,\text{m}^3)}{2 \times (206 \times 10^9\,\text{N/m}^2)}$$

$$= 148.4\,\text{N m (or J)}$$

(*Ans*)

3.2 Proof resilience and modulus of resilience

What is meant by the terms 'proof resilience' and 'modulus of resilience'?

Two bars are subjected to equal, gradually applied tensile loads. One bar is 50 mm diameter throughout and the other, which has the same length, is turned down to a diameter of 25 mm over the middle third of its length, the remainder having a diameter of 50 mm. Compare the strain energies of the two bars assuming that they are of the same material.

Compare also the amounts of energy that the two bars can absorb in simple tension without exceeding a given stress within the limit of proportionality.

Solution The *proof resilience* is the greatest amount of strain energy that can be absorbed by an elastic material and reappear when the loads are removed. If σ_p is the stress corresponding to the onset of permanent distortion,

$$\text{proof resilience} = \frac{\sigma_p{}^2}{2E} \times (\text{volume of the material})$$

In numerical examples σ_p may be taken as the proof stress or the limit of proportionality (see Volume 1, Chapter 9).

Modulus of resilience is defined as the proof resilience per unit volume and thus equals $\sigma_p{}^2/2E$. This quantity is a mechanical property of the material, e.g. for mild steel $E = 206\,\text{GN/m}^2$ and $\sigma_p = 300\,\text{MN/m}^2$ (approximately), and the modulus of resilience is about $220 \times 10^3\,\text{N m/m}^3$.

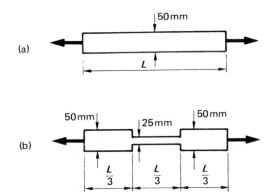

Figure 3.2

The two bars given in the question are shown in Figure 3.2(a) and (b). Suppose the length is L and that the tensile load applied to each is P.

Working in newton-metre units throughout, the strain energies are: For the uniform bar,

$$U = \left(\frac{P}{\frac{1}{4}\pi(0.050)^2}\right)^2 \times \frac{1}{2E} \times \tfrac{1}{4}\pi(0.050)^2 \times L$$
$$= 800LP^2/\pi E \tag{i}$$

For the end portions of the non-uniform bar which are 0.050 m diameter and length $\frac{2}{3}L$ (total) we have, from (i), replacing L by $\frac{2}{3}L$,

$$U = 1\,600LP^2/3\pi E \tag{ii}$$

For the middle portion of the non-uniform bar, which is 0.025 m diameter and length $\frac{1}{3}L$,

$$U = \left(\frac{P}{\frac{1}{4}\pi(0.025)^2}\right)^2 \times \frac{1}{2E} \times \tfrac{1}{4}\pi(0.025)^2 \times \tfrac{1}{3}L$$
$$= 3\,200LP^2/3\pi E \tag{iii}$$

If the load is the same for both bars, then, from (i), (ii) and (iii),

$$\frac{\text{Resilience of uniform bar}}{\text{Resilience of non-uniform bar}} = \frac{800LP^2/\pi E}{\dfrac{1\,600LP^2}{3\pi E} + \dfrac{3\,200LP^2}{3\pi E}}$$

$$= \tfrac{1}{2}$$

Suppose that, for the given stress, P is the load on the non-uniform bar. The maximum (given) stress occurs in the 25 mm diameter portion, and hence for the same stress to be attained in the uniform bar the load required is $(50/25)^2 \times P = 4P$ (the load being proportional to the square of the diameter). Replacing P by $4P$ in (i), the required ratio is

$$\frac{\text{Resilience of uniform bar}}{\text{Resilience of non-uniform bar}} = \frac{800L(4P)^2/\pi E}{\dfrac{1\,600LP^2}{3\pi E} + \dfrac{3\,200LP^2}{3\pi E}}$$

$$= 8$$

3.3 Maximum stress due to tensile impact load

A body whose weight is W falls a distance h before beginning to stretch a bar, length L, and cross-sectional area A. Derive expressions for the maximum stress induced in the bar when

(a) the maximum extension is negligible compared with h,
(b) the maximum extension is of the same order as h.

A bar, 3 m long and 50 mm diameter, hangs vertically and has a collar securely attached at the lower end. Find the maximum stress induced when

(i) a body of weight 2 kN falls 100 mm on to the collar
(ii) a body of weight 20 kN falls 10 mm on to the collar.

Take $E = 200 \text{ GN/m}^2$.

Solution Suppose the body falls the given distance on to a collar at the lower end of a vertical bar as shown in Figure 3.3. A small oscillation is set up and, provided the limit of proportionality is not exceeded, the collar will take up the same final position as in the case of a gradually applied load W. The maximum instantaneous extension is much greater than the steady value. Suppose it to be x as shown in the diagram.

Decrease in potential energy of the body
= strain energy of the bar

$$W(h + x) = \frac{\sigma^2}{2E} \times AL \tag{1}$$

where σ is the maximum stress induced.

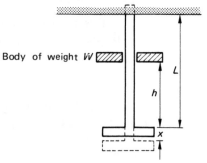

Body of weight W

Figure 3.3

(a) If x is negligible compared with h then, from (1),

$Wh = \sigma^2 AL/2E$

$$\sigma = \sqrt{\frac{2EWh}{AL}} \qquad\qquad (2)$$

(b) In this case x must be expressed in terms of σ,

$$x = \text{strain} \times L = \frac{\text{stress}}{E} \times L = \frac{\sigma L}{E}$$

Substituting in (1),

$$W\left(h + \frac{\sigma L}{E}\right) = \frac{\sigma^2 AL}{2E}$$

$$\left(\frac{AL}{2E}\right)\sigma^2 - \left(\frac{LW}{E}\right)\sigma - Wh = 0$$

Multiplying through by E/AL,

$$\tfrac{1}{2}\sigma^2 - \left(\frac{W}{A}\right)\sigma - \frac{WEh}{AL} = 0$$

This is a quadratic in σ and its solution is

$$\sigma = \frac{W}{A} \pm \sqrt{\left[\left(\frac{W}{A}\right)^2 + 4(\tfrac{1}{2})\left(\frac{WEh}{AL}\right)\right]}$$

$$= \frac{W}{A}\left(1 + \sqrt{\left[1 + \frac{2AEh}{WL}\right]}\right) \qquad\qquad (3)$$

the positive square root giving the maximum stress.

For the bar given in the question the maximum extension is probably less than 1/500 of the total length since the material is presumably steel with a limit of proportionality of 300–400 MN/m². Hence x is less than 6 mm (approx.).

(i) x is negligible compared with h and, therefore, using equation (2) the maximum stress is

$$\sigma = \sqrt{\frac{2 \times (200 \times 10^9 \text{ N/m}^2) \times (2 \times 10^3 \text{ N}) \times (0.1 \text{ m})}{\tfrac{1}{4}\pi \times (0.050 \text{ m})^2 \times 3 \text{ m}}}$$

$$= 116.5 \text{ MN/m}^2 \qquad\qquad (Ans)$$

(ii) Using the 'exact' formula (3), maximum stress is

$$\sigma = \frac{20 \times 10^3 \text{ N}}{\tfrac{1}{4}\pi \times (0.050) \text{ m}^2}$$

$$\times \left[1 + \sqrt{\left(1 + \frac{2 \times \tfrac{1}{4}\pi \times (0.050 \text{ m})^2 \times (200 \times 10^9 \text{ N/m}^2) \times 0.01 \text{ m}}{(20 \times 10^3 \text{ N}) \times 3 \text{ m}}\right)}\right]$$

$$= 127.1 \text{ MN/m}^2 \qquad\qquad (Ans)$$

Note Equation (2) would give the same answer for (i) and (ii).

3.4 Strain energy in shear

Derive an expression for the strain energy in a material subjected to a uniform shear stress.

Calculate the local strain energy at a point in a material where the shear stress is $80 \, \text{MN/m}^2$. $G = 28 \, \text{GN/m}^2$.

Solution Consider a rectangular block length l, width b, and height h as shown in Figure 3.4. Suppose that the block is rigidly fixed at its base and a shear force F is gradually applied along the top face as shown. If the resulting deflection in the direction of the force is x, the shear strain is $\gamma = x/h$. The strain energy is

U = work done by the shear force

= average force × deflection

$= \frac{1}{2}Fx$

Since $F = \tau \times bl$ (where τ = shear stress) and $x = \gamma h$ we have

$$U = \tfrac{1}{2}\tau bl \times \gamma h$$

$$= \tfrac{1}{2}\tau \times \frac{\tau}{G} \times blh$$

$$= \frac{\tau^2}{2G} \times (\text{volume of the block})$$

where G = modulus of rigidity = τ/γ.

The shear strain energy per unit volume is $\tau^2/2G$. If τ_p is the proof shear stress then $\tau_p^2/2G$ is called the *modulus of shear resilience*.

Substituting the values given in the question, the local strain energy per unit volume is

$$\frac{\tau^2}{2G} = \frac{(80 \times 10^6 \, \text{N/m}^2)^2}{2(28 \times 10^9 \, \text{N/m}^2)}$$

$$= 114.3 \times 10^3 \, \text{N m/m}^3 \qquad\qquad (Ans)$$

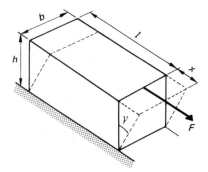

Figure 3.4

3.5 Strain energy in bending

Show that the strain energy of a beam or cantilever can be expressed as

$$\int_0^L \frac{M^2}{2EI}\, dx$$

and give the meaning of each symbol.

Compare the total strain energies of two equal uniform beams (simply-supported at their ends), one carrying a concentrated central load and one carrying a uniformly distributed load when

(a) the total load on each beam is the same,

(b) the maximum stress due to bending in each beam is the same.

Solution Suppose a bending couple M is gradually applied to the free end of a uniform cantilever, length L. The angular displacement it produces is equal to the slope of the deflected cantilever at this end and in Volume 1, Chapter 6, this is shown to be ML/EI, where E is the modulus of elasticity and I is the appropriate second moment of area for the cross-section. The strain energy is equal to the work done and hence:

$$U = \text{average moment (or couple)} \times \text{angular displacement}$$

$$= \tfrac{1}{2}M \times \frac{ML}{EI}$$

$$= \frac{M^2L}{2EI}$$

In the general case, M varies along the beam or cantilever and this result applies to a short length δx (instead of L). The strain energy of the whole beam or cantilever is therefore given by

$$U = \int_0^L \frac{M^2}{2EI}\, dx$$

as required, M being the bending moment at any distance x from the origin. L and I are defined above and E is Young's modulus.

Note The above theory takes no account of the strain energy due to shear.

For a simply-supported beam, with a central point load the bending moment at any distance x from one end is

$$M = x \times \text{reaction} = Wx/2$$

provided that x is less than $L/2$. Since the bending diagram is , symmetrical about mid-span and EI is constant for a uniform beam

$$U = 2 \int_0^{L/2} \frac{M^2}{2EI}\, dx = \frac{1}{EI} \int_0^{L/2} \left(\frac{Wx}{2}\right)^2 dx$$

$$= \frac{W^2}{4EI} \int_0^{L/2} x^2\, dx$$

$$= \frac{W^2}{4EI} \left[\frac{x^3}{3}\right]_0^{L/2}$$

$$= \frac{W^2 L^3}{96EI} \tag{i}$$

For a uniformly distributed load w/unit length the reaction at each end is $wL/2$ and the bending moment at any distance x from one end is

$$M = \tfrac{1}{2}wLx - \tfrac{1}{2}wx^2 = \tfrac{1}{2}w(Lx - x^2)$$

The total strain energy is thus

$$U = \int_0^L \frac{M^2}{2EI}\, dx = \frac{1}{2EI} \int_0^L \left[\frac{w}{2}(Lx - x^2)\right]^2 dx$$

$$= \frac{w^2}{8EI} \int_0^L (L^2 x^2 - 2Lx^3 + x^4)\, dx$$

$$= \frac{w^2}{8EI} \left[\frac{L^2 x^3}{3} - \frac{Lx^4}{2} + \frac{x^5}{5}\right]_0^L$$

$$= \frac{w^2 L^5}{240EI}$$

$$= \frac{W^2 L^3}{240EI} \tag{ii}$$

where $W = wL$, the total load.

(a) If the total load is the same for both beams, then from (i) and (ii), the required ratio is

$$\frac{\text{strain energy for point load case}}{\text{strain energy for distributed load case}} = \left(\frac{W^2 L^3}{96EI}\right) \bigg/ \left(\frac{W^2 L^3}{240EI}\right)$$

$$= 240/96 = 5/2 \qquad (Ans)$$

(b) Let W be the central point load. For the same maximum bending moment (and hence the same maximum bending stress) the distributed load is $2W$. Replacing W by $2W$ in (ii) the required ratio is

$$\frac{\text{strain energy for point load case}}{\text{strain energy for distributed load case}} = \left[\frac{W^2 L^3}{96EI}\right] \bigg/ \left[\frac{(2W)^2 L^3}{240EI}\right]$$

$$= 240/(96 \times 2^2) = 5/8 \qquad (Ans)$$

3.6 Deflection of steel ring by strain energy methods

A steel ring of rectangular cross-section 8 mm wide by 5 mm thick has a mean diameter of 300 mm. A narrow radial saw cut is made and tangential separating forces of 5 N each are applied at the cut in the plane of the ring.

Determine the additional separation due to these forces. Modulus of elasticity = 200 GN/m². (UL)

Solution Problems of this kind can be solved by equating the work done by the applied force to the bending strain energy of the ring. Let R = mean radius of the ring. Then, in the notation of Figure 3.5, the bending moment at X is given by:

$$M = P \times XY$$
$$= P \times (OB - OA)$$
$$= PR(1 - \cos \theta)$$

An expression for bending strain energy was derived in the previous solution in the form of an integral. In the present case it is necessary to change the variable from x to θ. Thus $dx = R \cdot d\theta$ and the limits of integration are now 0 and 2π. Hence:

$$U = \int_0^{2\pi} \frac{M^2 R}{2EI} d\theta$$

$$= \int_0^{2\pi} \frac{P^2 R^2 (1 - \cos \theta)^2 R}{2EI} d\theta$$

$$= \frac{P^2 R^3}{2EI} \int_0^{2\pi} (1 - 2\cos \theta + \cos^2 \theta)\, d\theta$$

$$= \frac{P^2 R^3}{2EI} \int_0^{2\pi} (\tfrac{3}{2} - 2\cos \theta + \tfrac{1}{2}\cos 2\theta)\, d\theta$$

[since $\cos^2 \theta = \tfrac{1}{2}(1 + \cos 2\theta)$]

$$= \frac{P^2 R^3}{2EI} [\tfrac{3}{2}\theta - 2\sin \theta + \tfrac{1}{4}\sin 2\theta]_0^{2\pi} = \frac{3\pi P^2 R^3}{2EI}$$

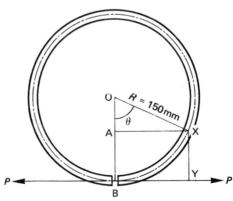

Figure 3.5

If x is the separation due to the forces P, the work done when they are applied is $\frac{1}{2}Px$ and, equating this to the last result, we have:

$$x = \frac{3\pi PR^3}{EI} \qquad \text{(i)}$$

and for a rectangular cross-section $b \times d$, $I = bd^3/12$. The cross-sectional dimensions given in the question can be interpreted in two ways. If the 8 mm width is taken to be in the radial direction, $b = 5$ mm and $d = 8$ mm. Thus:

$$I = \frac{bd^3}{12} = \frac{5 \times 8^3}{12} = 213.3 \text{ mm}^4$$

Substituting in (i) and working in mm units,

$$x = \frac{3\pi \times 5 \times 150^3}{200 \times 10^9 \times 10^{-6} \times 213.3}$$

$$= 3.73 \text{ mm} \qquad (Ans)$$

However, if b is taken to be 8 mm and $d = 5$ mm the result is 9.54 mm.

3.7 Maximum stress in a beam due to impact

When a load of 20 kN is gradually applied at a certain point in a beam it produces a deflection there of 15 mm and a maximum bending stress of 75 MN/m². Calculate the maximum value of the momentary stress produced when a mass whose weight is 5 kN falls through a height of 100 mm on to the beam at the same point.

Solution The problem can be solved by considering the amounts of energy absorbed by the beam in the static and impact cases. Suppose a mass whose weight is W falls a distance h on to the beam, as shown in Figure 3.6, causing a maximum instantaneous deflection δ at the point where it strikes. The decrease in potential energy of the falling mass is $W(h + \delta)$. Let W_E be the equivalent static load, i.e. the load that would cause the same deflection δ at the same point when gradually applied. The work done by W_E is $\frac{1}{2}W_E\delta$. Hence, equating these results,

$$W(h + \delta) = \tfrac{1}{2}W_E\delta \qquad \text{(i)}$$

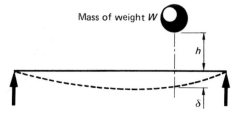

Mass of weight W

h

δ

Figure 3.6

It is usually necessary to express δ in terms of W_E. (In the case of central loading, for example, $\delta = W_E L^3/48EI$.) If this is done equation (i) becomes a quadratic for W_E. The solution can then be used in calculating any required stress or deflection using the usual equations for static loading.

Under static conditions the deflection and stress (but not the strain energy) are proportional to the load. Hence, with the figures given in the question and taking W_E in kN, the corresponding deflection is:

$$\delta = \frac{W_E}{20} \times 15$$

and, substituting in (i),

$$5\left(100 + \frac{15W_E}{20}\right) = \tfrac{1}{2}W_E \cdot \frac{15W_E}{20}$$

from which

$$\frac{1}{2}W_E^2 - 5W_E - \frac{2\,000}{3} = 0$$

This is a quadratic equation, the solution being

$$W_E = 5 \pm \sqrt{\left(25 + \frac{4\,000}{3}\right)} = 41.9$$

taking the positive root.

The maximum stress and deflection in the impact case are the same as for a static load of this amount. Hence, by proportion, the maximum stress produced by the falling load is

$$\sigma = \frac{41.9}{20} \times 75 \text{ MN/m}^2 = 157 \text{ MN/m}^2 \tag{Ans}$$

3.8 Strain energy in hollow shaft due to torsion

Derive an expression for the strain energy in a hollow shaft, external diameter D and internal diameter d, when it transmits a torque which produces a maximum shear stress τ.

A solid shaft, 150 mm diameter, is to be replaced by a hollow shaft of the same material, length, and weight. Calculate the diameters of the latter if its strain energy is to be 20 per cent greater than that of the solid shaft when transmitting torque at the same maximum shear stress.

Solution Suppose a torque T is gradually applied to the shaft and produces a twist θ (radians) in a length L. From the torsion equation $T/J = G\theta/L = \tau/r$ and, if τ is the maximum shear stress, $r = D/2$.

Thus

$$T = \frac{\tau}{r} J = \frac{2\tau}{D} \times \frac{\pi}{32} (D^4 - d^4) \tag{i}$$

$$\theta = \tau L / rG = 2\tau L / DG \tag{ii}$$

where G is the modulus of rigidity.

The strain energy for the shaft is, using (i) and (ii),

$U =$ work done by the torque

$\quad =$ average torque $\times$ angle of twist

$\quad = \frac{1}{2} T\theta$

$$\quad = \frac{1}{2} \left[\frac{2\tau}{D} \times \frac{\pi}{32} (D^4 - d^4) \right] \times \left[\frac{2\tau L}{DG} \right]$$

$$\quad = \frac{\tau^2}{16G} \times \frac{\pi (D^4 - d^4) L}{D^2}$$

$$\quad = \frac{\tau^2}{4G} \times \frac{(D^2 + d^2)}{D^2} \times \frac{\pi (D^2 - d^2) L}{4}$$

$$\quad = \frac{\tau^2}{4G} \times \frac{(D^2 + d^2)}{D^2} \times (\text{volume of the shaft}) \tag{iii}$$

Thus the *average* strain energy per unit volume is

$$\frac{\tau^2}{4G} \times \frac{(D^2 + d^2)}{D^2}$$

but the *local* strain energy per unit volume is not constant since the shear stress varies across the section. It was shown in Example 3.4 that for a *uniform* shear stress the strain energy is $(\tau^2/2G) \times$ volume. Hence the maximum value of $(D^2 + d^2)/D^2$ is 2, which corresponds to the case of a tube with an infinitely thin wall, so that $D = d$.

For a solid shaft, $d = 0$, and, from (iii), the total strain energy is

$$U = \frac{\tau^2}{4G} \times (\text{volume of the shaft}) \tag{iv}$$

Let D and d be the external and internal diameters of the required hollow shaft. Since the length, weight, and material is the same for the solid and hollow shafts then the volumes and cross-sectional areas of the two shafts are equal.

Hence, area of hollow shaft = area of solid shaft, or

$$\tfrac{1}{4}\pi(D^2 - d^2) = \tfrac{1}{4}\pi \times (150 \text{ mm})^2$$

$$D^2 - d^2 = 22\,500 \text{ mm}^2 \tag{v}$$

Also, for the required ratio of strain energies, we have, from (iii) and (iv),

Strain energy for hollow shaft = 1.2 × strain energy for solid shaft

$$\frac{\tau^2}{4G} \frac{(D^2 + d^2)}{D^2} \times \text{(volume)} = 1.2 \times \frac{\tau^2}{4G} \times \text{(volume)}$$

$$D^2 + d^2 = 1.2D^2$$

$$D^2 = 5d^2 \tag{vi}$$

Solving the simultaneous equations (v) and (vi),

$$d = 75 \quad \text{and} \quad D = 168$$

The required diameters are 168 mm external and 75 mm internal.

3.9 Strain energy due to perpendicular direct stresses

> If σ_1 and σ_2 are the principal stresses at a point in a material in a two-dimensional stress system, derive an expression for the strain energy, per unit volume, in terms of σ_1, σ_2, Poisson's ratio and Young's modulus.
>
> Find the strain energy in N m stored in a steel bar 250 mm long, and of cross-section 25 mm by 6 mm when it is subjected simultaneously to an axial pull of 18 kN and to a compressive stress of 60 MN/m² on its narrow edges.
>
> (For steel, $E = 200$ GN/m² and Poisson's ratio = 0.28.)
>
> *(IMechE)*

Solution From equation (ii) of Example 3.1, the strain energy per unit volume due to a single direct stress is

$$\frac{\sigma^2}{2E} = \tfrac{1}{2} \times \text{stress} \times \text{strain} \tag{i}$$

For simple stresses the strain is σ/E where $E =$ Young's modulus, but in complex stress systems the principal stresses and principal strains must be used. Hence, due to σ_1, the strain energy per unit volume is

$$\tfrac{1}{2} \times \sigma_1 \times \left(\frac{\sigma_1}{E} - \frac{v\sigma_2}{E}\right) = \frac{\sigma_1^2}{2E} - \frac{v\sigma_1\sigma_2}{2E} \tag{ii}$$

(where $v =$ Poisson's ratio), since the strain in the direction in which σ_1 acts is $\sigma_1/E - v\sigma_2/E$.

Similarly, the strain energy per unit volume due to σ_2 is

$$\frac{\sigma_2^2}{2E} - \frac{v\sigma_2\sigma_1}{2E} \tag{iii}$$

From (ii) and (iii), the total strain energy per unit volume is

$$\left(\frac{\sigma_1^2}{2E} - \frac{v\sigma_1\sigma_2}{2E}\right) + \left(\frac{\sigma_2^2}{2E} - \frac{v\sigma_2\sigma_1}{2E}\right) = \frac{1}{2E}(\sigma_1^2 + \sigma_2^2 - 2v\sigma_1\sigma_2) \tag{iv}$$

For the steel bar in the question, the principal stresses are

$$\sigma_1 \text{ (corresponding to the 18 kN load)} = \frac{18\,kN}{25\,mm \times 6\,mm}$$
$$= 120\,MN/m^2$$

and
$$\sigma_2 = -60\,MN/m^2$$

taking compressive stresses as negative.

From (iv), the total strain energy of the bar is

$$U = \frac{1}{2E}(\sigma_1^2 + \sigma_2^2 - 2v\sigma_1\sigma_2) \times \text{volume of the bar}$$

$$= \frac{1}{2 \times 200 \times 10^9}[(120 \times 10^6)^2 + (-60 \times 10^6)^2$$
$$- 2 \times 0.28 \times (120 \times 10^6)(-60 \times 10^6)]$$
$$\times [250 \times 25 \times 6 \times 10^{-9}]$$
$$= \frac{(60 \times 10^6)^2 \times 10^{-9}}{400 \times 10^9}$$
$$\times [2^2 + (-1)^2 - 2 \times 0.28 \times 2 \times (-1)] \times 37\,500$$
$$= 2.065\,N\,m \qquad\qquad (Ans)$$

3.10 Theories of failure

Explain briefly the various theories which have been put forward to obtain a criterion of failure under conditions of complex stress.

At a point in the wall of a thin steel tube there are perpendicular stresses of $40\,MN/m^2$ and $20\,MN/m^2$, both tensile. Calculate the equivalent stress in simple tension according to the maximum principal strain theory and the (total) strain energy theory.

Poisson's ratio $= 0.28$.

Solution Failure here means elastic breakdown and the onset of permanent strain. Let σ_Y be the stress at which this occurs in simple tension (for practical purposes the limit of proportionality may be used). Suppose in the three-dimensional complex stress system on the same material, the principal stresses are σ_1, σ_2 and σ_3 in descending order, tensions being positive.

The main theories of the conditions for elastic failure are as follows:

(a) *Greatest principal stress theory (Rankine)*. This states that failure occurs when the greatest principal stress reaches a critical value. For the complex stress case, this stress is σ_1. In simple tension it is σ_Y. Hence the criterion is

$$\sigma_1 = \sigma_Y \qquad\qquad (i)$$

(b) *Greatest principal strain theory (St Venant)*. This states that the relevant quantity is the greatest principal strain. In the complex stress case this is

$$\frac{\sigma_1}{E} - \frac{v\sigma_2}{E} - \frac{v\sigma_3}{E}$$

and in simple tension it is σ_Y/E. On equating these strains, we have

$$\sigma_1 - v(\sigma_2 + \sigma_3) = \sigma_Y \tag{ii}$$

(c) *Maximum shear stress theory (Coulomb, Tresca, Guest)*. The maximum shear stress on an interface is half the difference of the corresponding principal stresses. This is $\frac{1}{2}(\sigma_1 - \sigma_3)$ for the complex stress system and $\frac{1}{2}(\sigma_Y - 0)$ in simple tension. Thus, by this theory

$$\sigma_1 - \sigma_3 = \sigma_Y \tag{iii}$$

(d) *Total strain energy theory (Beltrami)*. By extending the theory given in Example 3.9 to three-dimensional conditions the total strain energy is

$$\frac{1}{2E}[\sigma_1{}^2 + \sigma_2{}^2 + \sigma_3{}^2 - 2v(\sigma_1\sigma_2 + \sigma_2\sigma_3 + \sigma_3\sigma_1)]$$

In simple tension the strain energy is $\sigma_Y{}^2/2E$ and this leads to the relationship

$$\sigma_1{}^2 + \sigma_2{}^2 + \sigma_3{}^2 - 2v(\sigma_1\sigma_2 + \sigma_2\sigma_3 + \sigma_3\sigma_1) = \sigma_Y{}^2 \tag{iv}$$

(e) *Mises–Hencky (shear strain energy) theory*. This states that the relevant quantity is

$$(\sigma_1 - \sigma_2)^2 + (\sigma_2 - \sigma_3)^2 + (\sigma_3 - \sigma_1)^2$$

and it can be shown (see Problem 24) that this expression represents the shear strain energy. In simple tension the principal stresses are σ_Y, 0 and 0, and the corresponding expression is therefore $2\sigma_Y{}^2$.
Hence the criterion is

$$(\sigma_1 - \sigma_2)^2 + (\sigma_2 - \sigma_3)^2 + (\sigma_3 - \sigma_1)^2 = 2\sigma_Y{}^2 \tag{v}$$

In the present example (working in MN/m^2) we have

$$\sigma_1 = 40 \qquad \sigma_2 = 20 \qquad \sigma_3 = 0$$

Principal strain theory. From (ii), the equivalent stress in simple tension is

$$\sigma = \sigma_1 - v(\sigma_2 + \sigma_3) = 40 - 0.28(20 + 0)$$
$$= 34 \text{ MN/m}^2 \tag{Ans}$$

Total strain energy theory. From (iv)

$$\sigma^2 = \sigma_1{}^2 + \sigma_2{}^2 + \sigma_3{}^2 - 2v(\sigma_1\sigma_2 + \sigma_2\sigma_3 + \sigma_3\sigma_1)$$
$$= 40^2 + 20^2 + 0 - 2 \times 0.28(40 \times 20 + 20 \times 0 + 0 \times 40)$$
$$= 1\,552$$
$$\sigma = 39.4 \text{ MN/m}^2 \tag{Ans}$$

3.11 Safety factors according to theories of failure

Discuss briefly the merits of the various theories of elastic failure.

A certain steel has a proportionality limit of $270 \, \text{MN/m}^2$ in simple tension. Under a certain two-dimensional stress system the principal stresses are $105 \, \text{MN/m}^2$ (tensile) and $30 \, \text{MN/m}^2$ (compressive). Calculate the factor of safety according to

(a) the maximum shear stress theory,
(b) the Mises–Hencky theory.

Solution The greatest principal stress theory is reasonably correct for brittle materials such as cast iron. The greatest principal strain theory is of very little value. The maximum shear stress theory is widely used for ductile materials, particularly in the design of shafts subjected to combined bending and torsion.

Some experimental results on ductile materials support the (total) strain energy theory but more are in agreement with the Mises–Hencky criterion, which is widely regarded as the most reliable basis for design.

Working in MN/m^2, the question gives

$$\sigma_1 = 105 \qquad \sigma_2 = 0 \qquad \sigma_3 = -30$$

(a) By the maximum shear stress theory, the equivalent single tensile stress is

$$\sigma = \sigma_1 - \sigma_3 = 105 - (-30) = 135$$

Factor of safety $= \sigma_Y / \sigma = 270/135 = 2$ (*Ans*)

(b) Mises–Hencky theory.

$$2\sigma^2 = (\sigma_1 - \sigma_2)^2 + (\sigma_2 - \sigma_3)^2 + (\sigma_3 - \sigma_1)^2$$
$$= (105 - 0)^2 + (0 + 30)^2 + (-30 - 105)^2$$
$$= 30\,150$$
$$\sigma = 122.8$$

Factor of safety $= \sigma_Y / \sigma = 270/122.8 = 2.2$ (*Ans*)

Note This result assumes that the factor of safety is defined as a ratio of stresses. For the total strain energy and Mises–Hencky (shear strain energy) theories it is sometimes regarded as the ratio of energies. In this case, answer (b) would become $\sigma_Y^2 / \sigma^2 = 270^2 / (\frac{1}{2} \times 30\,150) = 4.84$.

3.12 Equivalent bending moment according to theories of failure

Explain the term 'equivalent bending moment' as used in connection with shafting subjected to a bending moment combined with a twisting moment.

For a shaft of solid circular section subjected to a bending moment M combined with a twisting moment T, deduce the formula for the equivalent bending moment M_E in terms of M, T and if necessary, Poisson's ratio v, to correspond with each of the following hypotheses of elastic failure: (a) maximum principal stress; (b) maximum shearing stress; (c) maximum strain energy.

For a shaft 10 cm diameter made of steel for which the limiting stress in simple tension is 120 MN/m² draw to scale a graph for each of the above theories showing the limits within which combined values of M and T must occur according to each of the hypotheses. $v = 0.286$. (UL)

Solution The equivalent bending moment (M_E) is that which, if acting alone, would produce the same value of the quantity, used for the criterion of failure, as the given bending moment (M) and twisting moment (T) in combination.

For a circular shaft $J = 2I$ and, if d is the external diameter, the stresses at the surface are

bending stress $\sigma_x = My/I = Md/2I$

shear stress $\tau_{xy} = Tr/J = Td/2J = Td/4I$

This is a case of combined stress in two dimensions in which $\sigma_y = 0$. The principal stresses (σ_1 and σ_2) are the roots of the equation,

$$(\sigma - \sigma_x)(\sigma - \sigma_y) = \tau_{xy}{}^2$$

$$\sigma^2 - \sigma\left(\frac{Md}{2I}\right) - \left(\frac{Td}{4I}\right)^2 = 0$$

and the principal stresses are

$$\sigma_1 \text{ and } \sigma_2 = \frac{d}{4I}[M \pm \sqrt{(M^2 + T^2)}] \qquad\qquad \text{(i)}$$

The third principal stress is zero and falls between these two values. For a bending moment M_E acting alone the maximum direct stress is

$$\sigma = M_E y/I = M_E d/2I \qquad\qquad \text{(ii)}$$

(a) *Maximum principal stress.* For this criterion $\sigma = \sigma_1$ and

$$\frac{M_E d}{2I} = \frac{d}{4I}[M + \sqrt{(M^2 + T^2)}]$$

$$M_E = \tfrac{1}{2}[M + \sqrt{(M^2 + T^2)}]$$

(b) *Maximum shearing stress.* In the combined case the maximum shearing stress is $\tfrac{1}{2}(\sigma_1 - \sigma_2)$ and in the equivalent bending moment case it is $\tfrac{1}{2}(\sigma - 0)$. Thus from (i) and (ii)

$$\frac{1}{2}\left(\frac{M_E d}{2I}\right) = \frac{1}{2}\left(\frac{d}{4I}\right) \times \{[M + \sqrt{(M^2 + T^2)}] - [M - \sqrt{(M^2 + T^2)}]\}$$

$$M_E = \sqrt{(M^2 + T^2)}$$

(c) *Maximum strain energy.* Putting $\sigma_3 = 0$ in equation (iv) of Example 3.10 we have

$$\left(\frac{M_E d}{2I}\right)^2 = \left\{\frac{d}{4I}[M + \sqrt{(M^2 + T^2)}]\right\}^2 + \left\{\frac{d}{4I}[M - \sqrt{(M^2 + T^2)}]\right\}^2$$
$$- 2\nu\frac{d^2}{16I^2}[M + \sqrt{(M^2 + T^2)}][M - \sqrt{(M^2 + T^2)}]$$

from which

$$M_E^2 = M^2 + \tfrac{1}{2}T^2(1 + \nu)$$

With the numerical data given in the question

$$M_E = \sigma I/y = \sigma \times \pi d^3/32$$
$$= \tfrac{1}{32}[120 \text{ MN/m}^2 \times \pi \times (10 \text{ cm})^3] = 11.8 \text{ kN m}$$

Taking values of M from 0 to 11.8 kN m and using the three criteria mentioned the maximum values of T are as follows:

M	kN m	0	2	4	6	8	10	11.8
T kN m	(a)	23.6	21.5	19.2	16.5	13.4	9.2	0
	(b)	11.8	11.6	11.1	10.2	8.7	6.3	0
	(c)	14.7	14.5	13.8	12.7	10.8	7.9	0

Figure 3.7 illustrates the results and pairs of values of M and T must lie within each boundary to satisfy the corresponding criterion.

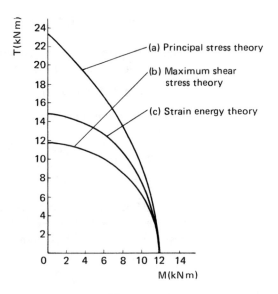

Figure 3.7

Problems

1 Find the total strain energy of a mild steel bar, 25 mm diameter and 2 m long, when it is subjected to a steady tensile load of 60 kN. $E = 200 \text{ GN/m}^2$.

Answer   36.7 N m (or J).

2 Two round bars have the same length L. One is diameter d for a length $L/3$ and diameter $2d$ for a length $2L/3$. The other is diameter $2d$ for a length $2L/3$ and diameter $3d$ for a length $L/3$. Compare the amounts of energy stored in the bars when they are subjected to equal tensile loads, assuming that they are of the same material.

Compare also the amounts of energy that can be absorbed by the two bars in simple tension without exceeding a given stress within the limit of proportionality.

Answer   For same load, $\dfrac{\text{resilience of first bar}}{\text{resilience of second bar}} = \dfrac{27}{11}$; for same stress, ratio is 27/176.

3 Calculate the modulus of resilience in tension and in shear for a steel having the following properties.

Limits of proportionality: in tension, 325 MN/m^2; in shear, 240 MN/m^2.

$E = 206 \text{ GN/m}^2$; $G = 80 \text{ GN/m}^2$.

Answer   In tension, 256 kN m/m^3; in shear, 360 kN m/m^3 (or kJ/m^3).

4 A solid shaft carries a flywheel of mass 150 kg and having a radius of gyration of 0.5 m. The shaft is rotating at a steady speed of 60 rev/min when it is suddenly fixed at a point 4 m from the flywheel. Calculate the shaft diameter if the maximum instantaneous shear stress produced is 150 MN/m^2. Assume that the kinetic energy of the flywheel is taken up as torsional strain energy by the shaft. Neglect the inertia of the shaft.

$G = 80 \text{ GN/m}^2$.

Answer   57.9 mm.

5 A bar, 1.2 m long and 25 mm diameter hangs vertically and has a collar rigidly attached at the lower end. When a mass of 500 kg is gradually lowered on to the collar, the extension is 0.05 mm. Find the maximum instantaneous stress induced in the bar when this mass falls a distance 2.5 mm on to the collar.

Answer 110.4 MN/m^2.

6 Find the maximum height from which a mass of 2 500 kg may fall on to a column 100 mm diameter and 600 mm high, if the stress is limited to 100 MN/m^2. $E = 200 \text{ GN/m}^2$. If you use a formula, prove it.

Answer   4.8 mm.

7 Derive, by strain energy methods, an expression for the deflection at the mid-point of a simply-supported beam carrying a central point load.

A simply-supported beam is of symmetrical section 250 mm deep ($I = 10\,000\,\text{cm}^4$) and spans 4 m. From what height can a mass of 2 000 kg fall on to the mid-point of the beam without causing a maximum instantaneous bending stress greater than 120 MN/m^2? $E = 200\,\text{GN/m}^2$.

Answer 9.26 mm.

8 A uniform beam, length L, is simply supported at its ends and carries a concentrated load W at a point whose distances from the ends are a and b. Show that the bending strain energy with this loading is $W^2a^2b^2/6LEI$. Use the result to obtain an expression for the deflection at the point where the load is applied.

Answer $Wa^2b^2/3LEI$.

9 A uniform bar is used as a beam or cantilever with an effective length L and it carries a total load W which may be concentrated at one point or uniformly distributed throughout its length. The total bending strain energy is given by W^2L^3/kEI, where k is a numerical coefficient that depends upon the type of loading and supports. Determine k for each of the following cases:

    (a) cantilever with a point load at the free end;
    (b) cantilever with a uniformly distributed load;
    (c) simply-supported beam with a central point load;
    (d) simply-supported beam with a uniformly distributed load;
    (e) built-in (encastré) beam with a central point load;
    (f) built-in beam with a uniformly distributed load.

Answer (a) 6; (b) 40; (c) 96; (d) 240; (e) 384; (f) 720.

10 Explain the meaning of the term 'strain energy' and state the type of problems for which the conception may be used. A cantilever has its section tapered so that the 2nd moment of area I varies from I_0 at the support to zero at the free end, the value of I at a distance x from the free end being I_0x/l where l is the cantilever length. A load W is applied at the free end. By how much does the free end displace, assuming the material to be homogeneous and of Young's modulus E? Strain energy methods should be used. (*RAeS*)

Answer $Wl^3/2EI_0$.

11 Prove that the strain energy stored in a bent beam is

$$\int (M^2/2EI)\,dx$$

where M is the bending moment at any section distant x from the origin, E is the modulus of elasticity of the material at the section, I is the relevant moment of inertia at the section.

A load of 30 kN is applied gradually to the centre of a simply supported beam of 3 m span. The section has a moment of inertia of 1 500 cm^4, and E may be taken as 200 GN/m^2. Find the amount of strain energy stored in the beam, and hence find the deflection under the load. (*IStructE*)
Answer 84.4 N m; 5.63 mm.

12 Obtain an expression for the resilience of a solid circular bar of diameter d and length L when subjected to a uniform tensile stress σ_t.

A bar is 60 mm diameter and 1 m long. Determine the stress produced in it by a mass of weight 1.8 kN falling 80 mm before commencing to stretch the bar. ($E = 200$ GN/m^2). (*IStructE*)
Answer $\pi\sigma_t^2 d^2 L/8E$; 143.4 MN/m^2.

13 Explain what you understand by the term 'strain energy', and derive from first principles an expression giving the strain energy in a tension bar of length L and cross-sectional area A when under a load F.

A tension bar 3 m long is made up of two parts: 1.8 m of its length has a cross-sectional area of 12 cm^2, while the remaining 1.2 m has a cross-sectional area of 24 cm^2. An axial load of 80 kN is gradually applied.

Find the total strain energy produced in the bar and compare this value with that obtained in a uniform bar of the same length *and having the same volume* when under the same load.
$E = 200$ GN/m^2. (*IStructE*)
Answer $\dfrac{F^2 L}{2AE}$; 32 N m; $\dfrac{\text{strain energy in given bar}}{\text{strain energy in uniform bar}} = \dfrac{28}{25}$.

14 A rod of material for which Young's modulus is E has a cross-sectional area a and a length l. Obtain an expression for the strain energy stored in it when elastically stressed by a gradually applied direct load P.

A bolt 500 mm long has a diameter of 125 mm for one-half of its length and of 112 mm for the other half. It may, on occasion, be subjected to a maximum tensile force of 1 MN. Determine the total extension of the bolt and the strain energy in each part. Young's modulus, 200 GN/m^2. (*IMechE*)
Answer $P^2 l/2aE$; extension $= 0.229$ mm; strain energy $= 50.9$ N m (125 mm diameter part) and 63.5 N m (112 mm diameter part).

15 Two shafts, one of steel and the other of phosphor-bronze, are of the same length and are subjected to equal torques. If the steel shaft is 10 mm diameter find the diameter of the phosphor-bronze shaft so that it will store the same amount of strain energy, per unit volume, as the steel shaft. Also determine the ratio of the maximum shear stresses induced in the two shafts. Take the modulus of rigidity for phosphor-bronze as 48 GN/m², and for steel as 80 GN/m². *(IMechE)*

Answer Required diameter, 10.8 mm; $\dfrac{\tau \text{ (steel)}}{\tau \text{ (bronze)}} = 1.265$.

16 A hollow shaft, subjected to a pure torque, attains a maximum shearing stress τ. Given that the strain energy stored per unit volume is $\tau^2/3G$, where G is the modulus of rigidity, calculate the ratio of the shaft diameters. Determine the actual diameters for such a shaft required to transmit 4 MN at 110 rev/min with uniform torque when the energy stored is 20 000 N m/m³ of material; take $G = 80$ MN/m².
Answer $\sqrt{3}$; 306 and 177 mm.

17 A hollow shaft 200 mm external diameter and 125 mm internal diameter transmits 1.5 MW at a speed of 150 rev/min. Calculate the shearing stress at the inner and outer surfaces of the shaft and the strain energy per metre length. G (steel) = 80×10^9 N/m². *(UL)*
Answer 44.9 and 71.8 MN/m²; 428 N m.

18 What is meant by 'resilience'?
A bar of steel, l m in length and d m square in cross-section, is subjected to a uniform bending moment (a) in a plane parallel to one of the sides of the bar and (b) in the plane of a diagonal. The maximum fibre stress is σ N/m² within the elastic range. Derive an expression of the elastic strain energy in each case. *(UL)*

Answer (a) $\dfrac{\sigma^2 d^2 l}{6E}$ N m; (b) $\dfrac{\sigma^2 d^2 l}{12E}$ N m

19 Compare the strain energy of a beam, simply supported at the ends and loaded with a uniformly distributed load, with that of the same beam centrally loaded and having the same value of the maximum bending stress. *(UL)*

Answer $\dfrac{U \text{ for point load}}{U \text{ for distributed load}} = \dfrac{5}{8}$

20 A steel bar 60 mm diameter is bent to the shape shown in Figure 3.8 and the lower end is firmly fixed in the ground in a vertical position. A load of 1 kN is applied at the free end. Calculate the vertical deflection at the free end. $E = 200$ GN/m². *(UL)*

Answer 18.0 mm.

21 A concentrated load W gradually applied to a horizontal beam, simply supported at the ends, produces a deflection y at the load point. If this load falls through a distance h, before making contact with the beam, find an expression, in terms of h

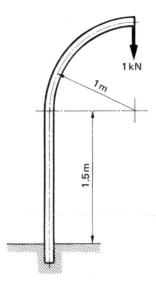

1 kN

1 m

1.5 m

Figure 3.8

and y, for the maximum deflection at the load point. Neglect loss of energy at impact.

In a given beam a concentrated load W, gradually applied, produces a deflection of 5 mm at the load point and a maximum bending stress of 60 MN/m². Find the greatest height from which a load $0.1W$ can be dropped without exceeding the elastic limit stress of 270 MN/m². *(UL)*

Answer   Maximum deflection $= y + \sqrt{(y^2 + 2yh)}$; 484 mm.

22 The shearing stress at the surface of a solid steel shaft is 75 MN/m² when subjected to the action of a torque. When used as a beam, simply supported at the ends and centrally loaded the greatest bending stress at the surface of the shaft is 120 MN/m². Compare the strain energies of the shaft for the two conditions of loading.

Is this ratio affected if the shaft is hollow and the stresses remain unchanged?

$(E = 210 \times 10^9 \,\text{N/m}^2; G = 80.5 \times 10^9 \,\text{N/m}^2).$ *(UL)*

Answer $\dfrac{\text{strain energy in torsion}}{\text{strain energy in bending}} = 6.11$; ratio is unaffected if shaft is hollow.

23   A hollow steel shaft has an external diameter of 225 mm and it was found that, when running at 125 rev/min and transmitting 2.2 MW, the angle of twist per metre length of shaft was 0.5°. If G is assumed to be $80 \times 10^9 \,\text{N/m}^2$, estimate the internal diameter of the shaft and the maximum shearing stress.

If the total strain energy U is expressed as $U = u \times$ volume of shaft, find the value of u; find also the strain energy per m³ of

material at the external surface of the shaft. (*UL*)
Answer 103 mm; 78.5 MN/m²; 23.3 and 38.6 kN m/m³.

24 By an extension of the method used in Example 3.9 show that at a point in a material subject to a three-dimensional stress system the strain energy, per unit volume, is

$$\frac{1}{2E}[\sigma_1{}^2 + \sigma_2{}^2 + \sigma_3{}^2 - 2v(\sigma_1\sigma_2 + \sigma_2\sigma_3 + \sigma_3\sigma_1)]$$

where σ_1, σ_2 and σ_3 are the principal stresses. Suppose σ is the mean of the three principal stresses, i.e. $\sigma = \frac{1}{3}(\sigma_1 + \sigma_2 + \sigma_3)$, so that the principal stresses can be replaced by two systems as follows:

(a) a stress σ in each direction which produces a change in volume but no change in shape.

(b) perpendicular stresses $(\sigma_1 - \sigma)$, $(\sigma_2 - \sigma)$ and $(\sigma_3 - \sigma)$ in the directions of σ_1, σ_2 and σ_3 which produce a change in shape but no change in volume.

Show that the strain energies due to these systems are respectively

(a) $\dfrac{3\sigma^2}{2E}(1 - 2v)$

(b) $\dfrac{1}{12G}[(\sigma_1 - \sigma_2)^2 + (\sigma_2 - \sigma_3)^2 + (\sigma_3 - \sigma_1)^2]$

Check that the sum of these results equals the expression found above.

25 A rolled steel beam is simply supported on a span of 6 m. A gradually applied load of 6 kN acting at mid-span produces a deflection of 1.08 mm at mid-span. Determine the value of *EI*, the flexural constant for the beam, stating the units in which it is expressed.

If the load of 6 kN is dropped on to the beam at mid-span, falling freely through a distance of 20 mm before striking the beam, determine the maximum deflection produced. (*UL*)
Answer 25 × 10⁶ N m²; 7.74 mm.

26 A vertical steel rod, of 25 mm diameter, checks the fall on its end of a weight of 2.25 kN which drops through a distance of 4 mm before it strikes the rod. Find the shortest length of rod which will bear the impact if the stress is not to exceed 136 MN/m². *E* = 200 GN/m².

Verify that the length found is the *least* possible length. (*UL*)
Answer 425 mm.

27 A beam of uniform section and span *l* is firmly fixed at the ends. Show that, due to a central load *P*, the maximum deflection is $Pl^3/192EI$.

A steel bar of rectangular section 40 mm wide and 10 mm deep is arranged in a horizontal position as a beam with firmly fixed ends and of span 0.9 m. A load of 140 N is allowed to fall freely on to the beam at mid-span. Find the height above the beam from which the load must fall in order to produce in the beam a maximum stress of 140 MN/m². Take $E = 200$ GN/m².

Describe briefly the behaviour of the beam immediately after impact. (UL)

Answer 9.28 mm.

28 A piece of steel is subjected to the following system of two-dimensional stress, namely, two tensile stresses of 90 and 45 MN/m² on planes at right angles to each other and a shear stress of 30 NM/m² on the same planes.

Working from first principles, find

(i) the principal stresses;
(ii) the principal strains;
(iii) the strain energy per unit volume.

If the tensile stress of 45 MN/m² is reversed in sign, find the new value to which the 90 MN/m² stress must be changed, assuming it to remain tensile, so that the strain energy remains the same, and find the new principal strains.

Take $E = 200$ GN/m², $G = 80$ GN/m², Poisson's ratio $= 0.25$ (UL)

Answer (i) 105 and 30 Mn/m² (tensile); (ii) 487×10^{-6} and 18.5×10^{-6}; (iii) 25.9 kN m/m³; stress must be changed to 67.5 MN/m² (tensile); 441×10^{-6} and -356×10^{-6}.

29 Show that the strain energy per unit volume in a two-dimensional system is given by the formula

$$U = \frac{1}{2E}(\sigma_1{}^2 + \sigma_2{}^2 - 2v\sigma_1\sigma_2)$$

in which σ_1 and σ_2 are the principal stresses. Hence show in the case of combined bending and torsion

$$U = \frac{1}{2E}(\sigma^2 + 2[1 + v]\tau^2)$$

in which σ is the maximum bending stress, τ the maximum shear stress and v is Poisson's ratio.

A 50 mm diameter mild steel shaft when subjected to pure torsion ceases to be elastic when the torque reaches 4.2 kN m. A similar shaft is subjected to a torque of 2.5 kN m and a bending moment M kN m. If maximum strain energy is the criterion of elastic failure find the value of M. Poisson's ratio $v = 0.28$. (IMechE)

Answer 27 kN m.

30 A horizontal circular shaft of a diameter d and diametral moment of inertia I is subjected to a bending moment $M \cos \theta$ in a vertical plane and to an axial twisting $M \sin \theta$. Show that the principal stresses at the ends of a vertical diameter are $\frac{1}{2}Mk(\cos \theta \pm 1)$, where $k = d/2I$.

If strain energy is the criterion of failure, show that

$$S = \frac{S_0 \sqrt{2}}{\sqrt{[\cos^2 \theta(1 - v) + (1 + v)]}}$$

where

S = maximum shearing stress,
S_0 = maximum shearing stress in the special case when $\theta = 0$,
v = Poisson's ratio. (*UL*)

31 Show that for a material subjected to two principal stresses, σ_1 and σ_2 the strain energy per unit volume of material is equal to

$$\frac{1}{2E}(\sigma_1{}^2 + \sigma_2{}^2 - 2v\sigma_1\sigma_2)$$

A thin-walled steel tube, of internal diameter 150 mm, closed at its ends, is subjected to an internal fluid pressure of 3 MN/m². Find the thickness of the tube if the criterion of failure is the maximum strain energy. Assume a factor of safety of 4 and take the elastic limit in pure tension as 300 MN/m². Poisson's ratio $v = 0.28$. (*IMechE*)
Answer 2.96 mm (assuming safety factor is ratio of stresses), 1.48 mm (assuming safety factor is ratio of energies).

32 For a certain material subjected to plane stress it is assumed that the criterion of elastic failure is the shear strain energy per unit volume. By considering coordinates relative to two axes at 45° to the principal axes show that the limiting values of the two principal stresses can be represented by an ellipse having semi-diameters $\sigma_e\sqrt{2}$ and $\sigma_e\sqrt{\frac{2}{3}}$ where σ_e is the equivalent simple tension. Hence show that for a given value of the major principal stress the elastic factor of safety is greatest when the minor principal stress is half the major, both stresses being of the same sign. (*UL*)

33 Derive a formula for the shear strain-energy in an element of material subjected to principal stresses σ_1 and σ_2 with the third principal stress zero.

A circular shaft 100 mm diameter is subjected to combined bending and twisting moments, the bending moment being three times the twisting moment.

If the direct tension yield-point of the material is 360 MN/m² and the factor of safety on yield is to be 4, calculate the allowable twisting moment by the three following theories of elastic failure:

(a) maximum principal stress theory,
(b) maximum shearing stress theory,
(c) maximum shear strain-energy theory.   (UL)

Answer   (a) 2.86, (b) 2.79, (c) 2.83 kN m.

34 A cast iron cylinder has outside and inside diameters of 20 cm and 12 cm. If the ultimate tensile strength of the cast iron is 150 MN/m^2 and its Poisson's ratio is 0.25, find, according to each of the following theories of failure, the internal pressure which would cause rupture:

(a) Maximum principal stress theory.
(b) Maximum principal strain theory.
(c) Maximum strain-energy theory.

Assume no longitudinal stress in the cylinder. Which of the results obtained do you consider should be applied to this case?
     (UL)

Answer   (a) 70.6 MN/m^2, (b) 63.2 MN/m^2, (c) 58.5 MN/m^2; (a) should be applied.

35 A sample of steel is tested (a) by direct tension of a solid bar and (b) by submitting a cantilevered circular tube to a load at the free end causing both torsion and bending. The limit of proportionality was found in case (a) to be a direct tensile stress of 262 MN/m^2 and in case (b) to be at a bending stress of 124 MN/m^2 together with a shearing stress of 117 MN/m^2.

Examine these results and state whether they are consistent with any of the following theories of elastic failure: (i) maximum principal stress theory, (ii) maximum shearing stress theory, (iii) maximum strain energy theory.

Assume reasonable values for any constants not given.   (UL)
Answer   Consistent with (ii).

36 A 50 cm diameter shaft is made of a material which in a direct tension test gave elastic failure at 340 MN/m^2. Poisson's ratio for the material is 0.3.

Estimate the torque which will just cause failure in the shaft when applied in addition to a bending moment of 3 500 N m taking as the criterion of failure

(i) the maximum principal stress,
(ii) the maximum strain energy.

Principal stress and strain energy formulae may be used without derivation; equivalent bending moment formulae may be used but they should be derived.   (UL)
Answer   (i) 3.35 kN m, (ii) 2.78 kN m.

37 A hollow circular shaft of external diameter D and internal diameter d transmits a torque T and has a maximum shear stress of τ_0 acting on its outer surface. Working from first principles derive an expression in terms of T and the angle of twist θ for the total strain energy stored in the shaft. Hence derive an expression in terms of τ_0, d, D and G for the mean strain energy per unit volume stored in the shaft.

Calculate the total strain energy per unit volume stored when $\tau_0 = 60\,\text{MN/m}^2$, $D = 75\,\text{mm}$, $d = 35\,\text{mm}$ and $G = 79\,\text{GN/m}^2$. If an equivalent solid circular shaft has the same shear stress acting on its outside surface when transmitting the same torque T, calculate the diameter of this solid shaft, and the average strain energy per unit volume stored within it.   (*Brunel*)

Answer $\frac{1}{2}T$; $\dfrac{\tau_0^2}{4G}\left(\dfrac{D^2 + d^2}{d^2}\right)$; 13.9 kN m/m^3 (or kJ/m^3); 73.8 mm; 11.4 kN m/m^3

38   At a point in a piece of elastic material there are three mutually perpendicular planes on which the stresses are as follows. On plane A there is a tensile stress of $480\,\text{MN/m}^2$ and a shear stress of $-160\,\text{MN/m}^2$. On plane B there is a compressive stress of $80\,\text{MN/m}^2$ and the complementary shear stress of $-160\,\text{MN/m}^2$ also acts. On plane C there is a tensile stress of $350\,\text{MN/m}^2$ and no shear stress. Find the magnitudes of the principal stresses and the positions of the planes on which they act.

A piece of the same material when tested in simple tension was found to yield at a stress of $610\,\text{MN/m}^2$. Using (a) the maximum shear stress theory and (b) the maximum shear strain energy theory, determine if yielding in the material, when subjected to the stress system described above, has occurred.   (*Brunel*)

Answer   522.5 MN/m^2 (tensile) and 122.5 MN/m^2 (compressive) on planes making angles $-14.9°$ and $75.1°$ respectively with A; 350 MN/m^2 (tensile) on C. (a) Yes, (b) No.

39   Working from first principles, derive the relationship between the elastic constants E, G and K.

The load on an 18 mm diameter bolt consists of an axial pull of 10 kN, together with a transverse shear force of 6 kN. If the modulus of rigidity for the bolt material is $80\,\text{GN/m}^2$ and Poisson's ratio $= 0.283$, calculate the total strain energy per unit volume stored in the bolt material at this point. The forces may be assumed to act uniformly over the bolt cross-section, and no other loads are applied.

If the elastic limit for the bolt material is $320\,\text{MN/m}^2$ and a factor of safety of 4 is to be imposed, calculate the minimum allowable bolt diameter according to (1) the maximum shear stress theory, and (2) the maximum strain energy of distortion theory.   (*Brunel*)

Answer $E = 9GK/(3K + G)$; $G = 3KE/(9K - E)$ or $K = GE/(9G - 3E)$; 724 kN m/m^3 (or kJ/m^3); (1) 70.5 mm; (2) 47.9 mm (if factor is ratio of stresses) or 33.9 mm (if factor is ratio of energies).

40 Show that the energy stored in an elastic body subject to the three principal stresses σ_1, σ_2 and σ_3 is given by:

$$U = \frac{1}{2E}[\sigma_1{}^2 + \sigma_2{}^2 + \sigma_3{}^2 - 2v(\sigma_1\sigma_2 + \sigma_2\sigma_3 + \sigma_3\sigma_1)]/\text{unit volume}$$

The principal stresses in a material, which fails elastically at 300 MN/m^2 when subject to uniaxial tensions, are such that $\sigma_1 = 2\sigma_2$ and $\sigma_3 = 0$. Poisson's ratio for the material is 0.3. Find the value of σ_1 corresponding to elastic failure using the total strain energy as a criterion. (*UL*)
Answer 308 MN/m^2.

41 A horizontal shaft of 8 cm diameter projects from a bearing and, in addition to the torque transmitted, the shaft carries a vertical load of 8 kN at 30 cm from the bearing. If the safe stress for the material, as determined in a simple tension test, is 150 MN/m^2, find the safe torque to which the shaft may be subjected using as the criterion (a) the maximum shearing stress, (b) the maximum strain energy. Poisson's ratio = 0.29.

A formlua for the principal stress may be used without proof but if a formula for either equivalent torque or equivalent bending moment is used such formula should be derived. (*UL*)
Answer (a) 7.15 N m; (b) 8.9 kN m.

4

Springs

Types of spring

A spring is a device in which the material is so arranged that it can withstand large distortions without the elastic limit being exceeded. Although straight bars in bending, torsion or tension are used as springs, there are two groups that are especially useful in practice; those consisting of a length of wire wound into a helix or spiral and those consisting of several rectangular plates strapped together.

Stiffness

The stiffness of a spring is defined as the load per unit deflection (or torque per unit angle of twist).

Proof load

This is the greatest load the spring can carry without suffering permanent distortion. The corresponding values of the maximum stress and the strain energy are called the proof stress and proof resilience respectively.

Close-coiled helical spring of circular wire with axial load W

If D = mean coil diameter, d = wire diameter and N = number of effective or 'free' coils the deflection and maximum shear stress are:

$$\delta = \frac{8WND^3}{Gd^4} \quad \text{and} \quad \tau_{max} = \frac{8WD}{\pi d^3}$$

where G = modulus of rigidity

Close-coiled helical spring with axial couple M

If L = total length of wire, I = relevant second moment of area of the wire section and Z = corresponding section modulus, the angular displacement (in radians) and maximum bending stress are:

$$\phi = \frac{ML}{EI} \quad \text{and} \quad \sigma_{max} = \frac{M}{Z}$$

where E = modulus of elasticity.

Flat spiral spring with axial couple M

The rotation of the spindle and maximum bending stress are:

$$\phi = \frac{ML}{EI} \quad \text{and} \quad \sigma_{max} = \frac{2M}{Z}$$

| Leaf, laminated or carriage spring | This consists of a number of plates or leaves strapped together, the lengths of the various plates varying so that the effective number at a given section is approximately proportional to the bending moment there.
The plates are initially curved and the spring is so arranged that they tend to straighten under the applied loads. The initial radius to which the plates are curved is generally chosen so that they become straight at the proof load.
The springs are often called elliptic although the curvature of the plates is normally circular. |
|---|---|

Semi-elliptic or beam type

The deflection and maximum stress due to a central load W are

$$\delta = \frac{3WL^3}{8Enbt^3} \quad \text{and} \quad \sigma_{max} = \frac{3WL}{2nbt^2}$$

Quarter-elliptic or cantilever type

The deflection and maximum stress due to an end load W are

$$\delta = \frac{6WL^3}{Enbt^3} \quad \text{and} \quad \sigma_{max} = \frac{6WL}{nbt^2}$$

where L = length of the spring,
$\quad b$ = breadth of each plate,
$\quad t$ = thickness of each plate,
$\quad n$ = number of plates,
$\quad E$ = Young's modulus

Worked examples

4.1 Shear stress in helical spring with axial load

> Derive an expression for the maximum shear stress in a close-coiled helical spring, mean coil diameter D, wire diameter d, subjected to an axial load W.
> A close-coiled helical spring is to carry a load of 100 N and the mean coil diameter is to be eight times the wire diameter. Calculate these diameters if the maximum shear stress is to be 100 MN/m².

Solution Figure 4.1(a) shows a helical (or cylindrical spiral) spring in which the coils are regarded as being very flat. At Figure 4.1(b) one coil is shown. The load W produces a twisting moment at any cross-section (e.g. at A) of (approximately)

$$T = W \times \text{radius of coils} = \tfrac{1}{2}WD \tag{i}$$

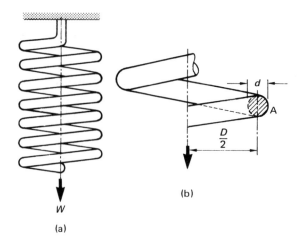

Figure 4.1

Assuming that the torsion equation applies at a section such as A the maximum shear stress is given by

$$\tau_{max} = \frac{T}{J} \times r = \frac{\frac{1}{2}WD \times \frac{1}{2}d}{\frac{1}{32}\pi d^4}$$

$$= \frac{8WD}{\pi d^3} \qquad \text{(ii)}$$

(Since the torsion equation was established for straight bars, equation (ii) is approximate only, the magnitude of the error depending on the ratio D/d.)

With the values given in the question,

$$\tau_{max} = 100 \text{ MN/m}^2, \qquad W = 100 \text{ N} \quad \text{and} \quad D = 8d$$

Thus, from (ii),

$$100 \times 10^6 = 8 \times 100 \times 8d/\pi d^3$$

$$d^2 = \frac{64}{\pi} \times 10^{-6} \qquad d = 4.51 \times 10^{-3} \text{ m}$$

Wire diameter $= 4.51$ mm   *(Ans)*
Mean coil diameter $= 8 \times 4.51 = 36.1$ mm   *(Ans)*

4.2 Deflection of concentric helical springs with axial load

Assuming without proof the formulae for the strength and stiffness of a round bar in torsion, derive an expression for the extension of a close-coiled helical spring in terms of the load W, the mean coil diameter D, the diameter of the wire section d, the number of active coils N, and the modulus of rigidity G.

The following data apply to two close-coiled helical steel springs.

Spring	N	D	d	Axial length uncompressed
A	8	100 mm	6 mm	70 mm
B	10	75 mm	5 mm	80mm

Spring B is placed inside A and both are compressed between a pair of parallel plates until the distance between the plates measures 60 mm. Calculate (a) the load applied to the plates, and (b) the maximum intensity of shear stress in each spring. $G = 80 \times 10^9 \, \text{N/m}^2$. (IMechE)

Solution Referring to equation (i) of Example 4.1, the twisting moment at every section of the wire is $WD/2$.

Assuming the torsion equation $T/J = G\theta/L = \tau/r$, the twist in a small length δL of the wire at this section is

$$\delta\theta = \frac{T\delta L}{GJ} = \frac{WD\delta L}{2GJ} \tag{i}$$

Consider the weight to be at the end of a radial arm; the distance it moves, due to this twist, is

$$\delta\theta \times \text{radius of the coil} = \frac{WD\delta L}{2GJ} \times \frac{D}{2}$$

$$= \frac{WD^2\delta L}{4GJ} \tag{ii}$$

Since the conditions are the same for every section of the wire, the total deflection of the spring is, from (ii),

$$\delta = \frac{WD^2L}{4GJ} \tag{iii}$$

where $L = $ total length of wire in the coil, and δ is now the symbol for deflection.

If N is the number of active (or 'free') coils the effective length of wire is

$$L = N \times \text{circumference of the coils}$$
$$= \pi ND \tag{iv}$$

Substituting from (iv) in (iii) and putting $J = \pi d^4/32$, the total deflection is

$$\delta = \frac{WD^2(\pi ND)}{4G(\pi d^4/32)} = \frac{8WND^3}{Gd^4} \tag{v}$$

Rearranging (v), $W = Gd^4\delta/8ND^3$ and since springs A and B are compressed 10 mm and 20 mm respectively, the total load on the parallel plates is (working in newtons and metres)

$$W = \left(\frac{Gd^4\delta}{8ND^3}\right) \text{ for A} + \left(\frac{Gd^4\delta}{8ND^3}\right) \text{ for B}$$

$$= \frac{80 \times 10^9}{8} \left[\frac{(0.006)^4 \times (0.01)}{8 \times (0.1)^3} + \frac{(0.005)^4 \times (0.02)}{10 \times (0.075)^3}\right]$$

$$= 45.9 \text{ N} \qquad\qquad (Ans)$$

Substituting the value of W found from (v) above in equation (ii) of Example 4.1, the maximum shear stress for each spring is

$$\tau_{max} = \frac{8D}{\pi d^3}\left(\frac{Gd^4\delta}{8ND^3}\right) = \frac{Gd\delta}{\pi ND^2} \qquad\qquad \text{(vi)}$$

Using (vi) the maximum shear stresses are

For spring A,

$$\tau_{max} = \frac{(80 \times 10^9) \times 0.006 \times 0.01}{\pi \times 8 \times (0.1)^2}$$

$$= 19.1 \text{ MN/m}^2 \qquad\qquad (Ans)$$

For spring B,

$$\tau_{max} = \frac{(80 \times 10^9) \times 0.005 \times 0.02}{\pi \times 10 \times (0.075)^2}$$

$$= 45.3 \text{ MN/m}^2 \qquad\qquad (Ans)$$

4.3 Axial deflection and angular twist of helical spring by strain energy methods

Use strain energy methods to find for a close-coiled helical spring

(a) the deflection due to an axial load W, and
(b) the angular twist due to an axial couple M. The spring is made of wire diameter d and the mean coil diameter is D.

Find the mass of such a spring which would absorb the energy of a truck of mass 10 tonne and moving at 1 m/s if

(i) the spring is compressed by the impact,
(ii) the spring is 'wound up' by the impact.

Working stresses: 350 MN/m² (bending) and 280 MN/m² (torsion). $E = 200 \times 10^9$ N/m² and $G = 80 \times 10^9$ N/m². The density of the spring material is 7.83 Mg/m³.

Solution (a) The twisting moment at every section of the wire is $T = WD/2$ as shown in Example 4.1. Also, the total angle of twist for the whole spring is, from the torsion equation,

$$\theta = \frac{TL}{GJ} = \frac{\pi TND}{G(\pi d^4/32)} = \frac{32TND}{Gd^4} \qquad\qquad \text{(i)}$$

since $L = \pi ND$, N being the number of coils.

Using (i), the total strain energy of the spring is

$$U = \tfrac{1}{2}T\theta = \tfrac{1}{2}T\left(\frac{32TND}{Gd^4}\right) = \frac{16T^2ND}{Gd^4} \qquad \text{(ii)}$$

Substituting for T and equating the work done by W to the strain energy

$$\tfrac{1}{2}W\delta = \frac{16(\tfrac{1}{2}WD)^2ND}{Gd^4}$$

and the deflection is

$$\delta = \frac{8WND^3}{Gd^4} \qquad \text{(iii)}$$

(b) At every section of the wire there is a bending moment equal to the applied couple M. Assuming that simple bending theory is applicable, the strain energy is (since M is a constant)

$$U = \frac{M^2L}{2EI} \qquad \text{(iv)}$$

where L is the total length of wire.

Equating the work done by M to the bending strain energy and substituting for L and I,

$$\tfrac{1}{2}M\phi = \frac{M^2L}{2EI}$$

and the angle of twist of one end of the spring relative to the other is

$$\phi = \frac{ML}{EI} = \frac{M(\pi ND)}{E(\pi d^4/64)} = \frac{64MND}{Ed^4} \qquad \text{(v)}$$

For the truck in question, the kinetic energy to be absorbed is given by

$$\text{KE} = \tfrac{1}{2}mv^2 \qquad \text{(vi)}$$

where m = mass of truck,
$\quad\quad v$ = velocity of truck.

Since $m = 10$ tonne $= 10 \times 10^3$ kg, we have from (vi)

$$\text{KE} = \tfrac{1}{2} \times (10 \times 10^3 \text{ kg}) \times (1 \text{ m/s})^2$$
$$= 5\,000 \text{ kg m}^2/\text{s}^2$$
$$= 5\,000 \text{ N m (or J)}$$

(i) If the spring absorbs the energy in direct compression then the 'wire' of the spring is in torsion. Hence from equation (iv), Example 3.8 of Chapter 3 the strain energy is

$$U = \frac{\tau_{max}^2}{4G} \times \text{volume of the spring}$$

Equating the strain energy to the kinetic energy and substituting for

τ_{max} (the maximum shear stress) and G,

$$\frac{\tau_{max}^2}{4G} \times (\text{volume of the spring}) = \text{KE of truck}$$

and volume of the spring is

$$\frac{(\text{KE of truck}) \times 4G}{\tau_{max}^2}$$

$$= \frac{5\,000\,\text{N m} \times 4 \times (80 \times 10^9\,\text{N/m}^2)}{(280 \times 10^6\,\text{N/m})^2}$$

$$= 0.0204\,\text{m}^3$$

Hence, the mass of the spring is

$$\text{Volume} \times \text{density} = 0.0204\,\text{m}^3 \times 7.83\,\text{Mg/m}^3$$

$$= 0.160 \times 10^6\,\text{Mg} = 160\,\text{kg} \qquad (Ans)$$

(ii) If σ_{max} is the maximum bending stress in the wire, $M = \sigma_{max}I/y$. Substituting in (iv), the strain energy is

$$U = \frac{(\sigma_{max}I/y)^2 L}{2EI} = \frac{\sigma_{max}^2 I L}{2y^2 E} = \frac{\sigma_{max}^2 (\pi d^4/64) L}{2(\frac{1}{2}d)^2 E}$$

$$= \frac{\sigma_{max}^2}{8E} \times (\tfrac{1}{4}\pi d^2 L)$$

$$= \frac{\sigma_{max}^2}{8E} \times (\text{volume of the spring}) \qquad (vii)$$

Equating this strain energy to the kinetic energy,

$$\frac{\sigma_{max}^2}{8E} \times (\text{volume of the spring}) = \text{KE of truck}$$

and the mass of the spring is

volume × density

$$= \left[\frac{\text{KE of truck} \times 8E}{\sigma_{max}^2}\right] \times \text{density}$$

$$= \frac{(5\,000\,\text{N m}) \times 8 \times (200 \times 10^9\,\text{N/m}^2)}{(350 \times 10^6\,\text{N/m}^2)^2} \times 7.83\,\text{Mg/m}^3$$

$$= 511\,\text{kg} \qquad (Ans)$$

4.4 Determination of number of plates for centrally loaded laminated spring

Derive an expression for the maximum bending stress in a semi-elliptic laminated spring consisting of n plates, each breadth b and thickness t, when it is subjected to a central load W. Find also the strain energy stored in the spring in terms of its volume.

How many plates, each 65 mm wide and 6 mm thick, would be required for such a spring 1 m long, if it is to carry a central load of 1.7 kN, the maximum bending stress being 140 MN/m²?

Solution A semi-elliptic laminated spring is shown diagrammatically in Figure 4.2. It is also known as a leaf or carriage spring and consists of a number of plates or leaves strapped together. Assuming that each plate is free to slide relative to its neighbours (i.e. neglecting friction between the plates) then, at any section, each plate acts as a separate beam.

Each plate overlaps its neighbour by an amount $L/2n$ at each end. Hence, a section XX, where there are x complete plates, is a distance $xL/2n$ from the end of the spring. The bending moment at XX is, therefore,

$$\frac{W}{2} \times \frac{xL}{2n} = \frac{WLx}{4n} \tag{i}$$

Since all the plates at XX have the same section and are constrained to bend to the same radius of curvature then the bending moment is shared equally between them. The bending moment for each plate at XX is

$$M = \frac{1}{x} \times \frac{WLx}{4n} = \frac{WL}{4n} \tag{ii}$$

Since the bending moment (ii) is independent of x it applies to all sections such as XX. In practice, to make the bending moment more uniform everywhere, the ends of the separate plates (except the longest) are often cut away as shown in Figure 4.2.

The maximum stress in each plate is, from the bending equation,

$$\sigma_{max} = \frac{M}{I} \times y_{max}$$

The bending moment is given by (ii), $I = bt^3/12$ and $y_{max} = \frac{1}{2}t$. Thus the maximum bending stress is

$$\sigma_{max} = \frac{(WL/4n)}{(bt^3/12)} \times \tfrac{1}{2}t = \frac{3WL}{2nbt^2} \tag{iii}$$

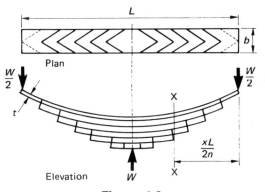

Figure 4.2

The strain energy stored in a beam carrying a uniform bending moment M is $(M^2/2EI) \times$ (length of the beam). For the given spring, substituting from (ii), the total strain energy is

$$U = \left(\frac{WL}{4n}\right)^2 \times \frac{1}{2EI} \times \text{(total length of all the leaves)}$$

$$= \left(\frac{\sigma_{max}bt^2}{6}\right)^2 \times \frac{1}{2EI} \times \text{(total length of all the leaves)}$$

since, from (iii), $WL/2n = \frac{1}{3}\sigma_{max}bt^2$.

$$U = \frac{(\sigma_{max}^2)b^2t^4}{36} \times \frac{1}{2E(bt^3/12)} \times \text{(total length of leaves)}$$

$$= \frac{\sigma_{max}^2}{6E} \times bt \times \text{(total length of leaves)}$$

$$= \frac{\sigma_{max}^2}{6E} \times \text{(volume of the spring)} \tag{iv}$$

For the values given in the question, the number of plates required is, from (iii),

$$n = \frac{3WL}{2\sigma_{max}bt^2}$$

$$= \frac{3 \times 1\,700\,\text{N} \times 1\,\text{m}}{2 \times (140 \times 10^6\,\text{N/m}^2) \times 0.065\,\text{m} \times (0.006\,\text{m})^2}$$

$$= 7.8$$

Taking the next highest whole number, 8 plates are required. *(Ans)*

4.5 Deflection of centrally loaded laminated spring

> A laminated spring, 900 mm long, is made up of plates each 50 mm wide and 6 mm thick. If the bending stress in the plates is limited to 80 MN/m², how many plates are required in order that the spring may carry a central point load of 800 N? If $E = 200 \times 10^9\,\text{N/m}^2$, what is the deflection under this load? Deduce the formula used for deflection.

Solution With the assumptions and notation of the previous solution there is a bending moment at each section of each plate of $M = WL/4n$. By the methods of Vol. 1, Chapter 6, the mid-span deflection of a uniform beam with a bending moment that is constant along its length is $ML^2/8EI$. Applying these results to the longest plate, the mid-span deflection of the spring is:

$$\delta = \frac{ML^2}{8EI} = \frac{WL}{4n} \cdot \frac{L^2}{8EI} = \frac{WL^3}{32nEI}$$

where I is the second moment of area for *one* plate. Putting $I = bt^3/12$, the result becomes

$$\delta = \frac{WL^3}{32nE(bt^3/12)} = \frac{3WL^3}{8nEbt^3} \tag{i}$$

For the spring given in the question, the number of plates required is (by equation (iii) of Example 4.4),

$$n = \frac{3WL}{2\sigma_{max}bt^2}$$

$$= \frac{3 \times (800\,\text{N}) \times (0.9\,\text{m})}{2 \times (80 \times 10^6\,\text{N/m}^2) \times (0.05\,\text{m}) \times (0.006\,\text{m})^2}$$

$$= 7.5$$

Taking the next highest whole number, 8 plates are required. (*Ans*)
In calculating the deflection, n must be taken as 8.
From (iv), the deflection under the given load is

$$\delta = \frac{3 \times (800\,\text{N}) \times (0.9\,\text{m})^3}{8 \times 8 \times (200 \times 10^9\,\text{N/m}^2) \times (0.05\,\text{m}) \times (0.006\,\text{m})^3}$$

$$= 12.7\,\text{mm} \tag{*Ans*}$$

4.6 Impact load on laminated spring

A leaf spring of the semi-elliptic type has 10 plates each 75 mm wide and 10 mm thick. The length of the spring is 1.25 m and the plates are of steel having a proof stress (bending) of 600 MN/m². to what radius should the plates be initially bent?

From what height can a mass of 45 kg fall on to the centre of the spring if the maximum stress produced is to be one-half of the proof stress?

$E = 200\,\text{GN/m}^2.$

Solution Leaf springs are usually given an initial curvature opposite to that produced by the load. The radius to which the plates are bent is so chosen that the spring just straightens under the proof load. In the present example, applying the bending equation to one plate, the required radius is

$$R = \frac{Ey}{\sigma} = \frac{E}{\sigma_{max}}\left(\frac{t}{2}\right)$$

$$= \frac{(200 \times 10^9\,\text{N/m}^2) \times (0.005\,\text{m})}{(600 \times 10^6\,\text{N/m}^2)}$$

$$= 1.67\,\text{m} \tag{*Ans*}$$

Let W_E be the equivalent static load which would produce the same maximum stress and deflection as the falling mass. In the usual notation,

$$\sigma_{max} = \frac{3W_E L}{2nbt^2}$$

$$W_E = \frac{2nbt^2 \sigma_{max}}{3L}$$

$$= \frac{2 \times 10 \times (0.075\,\text{m}) \times (0.01\,\text{m})^2 \times (300 \times 10^6\,\text{N/m}^2)}{3 \times (1.25\,\text{m})}$$

$$= 12\,\text{kN}$$

The corresponding deflection is

$$\delta = \frac{3W_E L^3}{8nEbt^3}$$

$$= \frac{3 \times (12 \times 10^3\,\text{N}) \times (1.25\,\text{m})^3}{8 \times 10 \times (200 \times 10^9\,\text{N/m}^2) \times (0.075\,\text{m}) \times (0.01\,\text{m})^3}$$

$$= 58.6\,\text{mm}$$

The weight of the falling mass is

$$W = mg = 45\,\text{kg} \times 9.81\,\text{m/s}^2 = 441\,\text{N}$$

Equating the decrease of potential energy of the falling mass to the strain energy of the spring (as in Example 3.7, Chapter 3)

$$W(h + \delta) = \tfrac{1}{2}W_E \delta$$

Hence the height from which the mass may fall is

$$h = \delta\left(\frac{W_E}{2W} - 1\right) = 58.6\,\text{mm}\left(\frac{12\,000}{2 \times 441} - 1\right)$$

$$= 738\,\text{mm} \hspace{4cm} (\textit{Ans})$$

4.7 Determination of plate sizes for laminated cantilever spring

> A laminated spring of the 'quarter-elliptic' type, 900 mm long, is to be made of steel plates which are available in multiples of 1 mm for thickness and 5 mm for width. The spring is to carry a load of 7.5 kN and the end deflection must not exceed 100 mm. The bending stress must not be greater than 300 MN/m².
>
> Find suitable values for the size and number of plates to be used, taking the width as five times the thickness.
>
> $E = 200\,\text{GN/m}^2$.

Solution A quarter-elliptic or cantilever laminated spring, length L, is shown diagrammatically in Figure 4.3. The load W is carried at the free end and it is clear that the maximum stress and deflection are equal to those of a semi-elliptic spring, length $2L$, carrying a central load $2W$ (and a reaction W at each end).

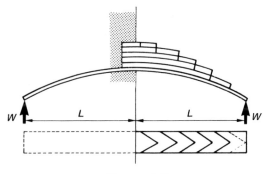

Figure 4.3

Hence, replacing W by $2W$ and L by $2L$ in equation (iii) of Example 4.4, the maximum stress for the quarter-elliptic spring is

$$\sigma_{max} = \frac{3(2W)(2L)}{2nbt^2} = \frac{6WL}{nbt^2} \tag{i}$$

Making a similar adjustment to equation (i) of Example 4.5, the deflection is

$$\delta = \frac{3(2W)(2L)^3}{8nEbt^3} = \frac{6WL^3}{nEbt^3} \tag{ii}$$

For the spring given in the question, we have, dividing (i) by (ii),

$$\frac{\sigma_{max}}{\delta} = \frac{6WL}{nbt^2} \Big/ \frac{6WL^3}{nEbt^3} = \frac{Et}{L^2}$$

or

$$t = \frac{\sigma_{max}L^2}{\delta E} = \frac{(300 \times 10^6 \,\text{N/m}^2) \times (0.9 \,\text{m})^2}{(0.1 \,\text{m}) \times (200 \times 10^9 \,\text{N/m}^2)}$$

$$= 12.15 \,\text{mm} \tag{iii}$$

From (iii), the size of plate required is 13 mm thick and 65 mm wide. Using this size, the number of plates required is, from (i),

$$n = \frac{6WL}{\sigma_{max}bt^2} = \frac{6 \times 7\,500 \,\text{N} \times 0.9 \,\text{m}}{(300 \times 10^6 \,\text{N/m}^2) \times 0.065 \,\text{m} \times (0.013 \,\text{m})^2}$$

$$= 12.25 \tag{iv}$$

Also, from (ii), the number is

$$n = \frac{6WL^3}{\delta Ebt^3}$$

$$= \frac{6 \times 7\,500 \,\text{N} \times (0.9 \,\text{m})^3}{0.1 \,\text{m} \times (200 \times 10^9 \,\text{N/m}^2) \times 0.065 \,\text{m} \times (0.013 \,\text{m})^3}$$

$$= 11.5 \tag{v}$$

From results (iv) and (v) the number of plates required to satisfy both conditions is 13. It is possible in certain problems, where the number of plates is small and the multiples of plate sizes are small, that in taking the next highest whole number for n the plate size can be reduced without the maximum deflection and stress being exceeded. In the present case, however, it will be found that 13 plates, each 60 mm × 12 mm, will not satisfy the conditions of the problem. Hence, 13 plates, each 65 mm × 13 mm, are required.

4.8 Angular twist and energy stored in flat spiral spring

A flat spiral spring has a total length of wire L. At the centre it is attached to a spindle and the other end is pinned. Obtain an expression for the angle through which the spindle turns when a couple M is applied to it. Find also the strain energy of the spring in terms of its volume and the maximum bending stress when it is made of wire of rectangular cross-section.

Calculate the length of steel strip 25 mm broad and 1 mm thick required for such a spring if one complete turn of the spindle is to produce a maximum bending stress of $150 \, \text{MN/m}^2$. $E = 200 \, \text{GN/m}^2$.

What is the strain energy of the spring for this condition?

Solution Figure 4.4 shows a flat spiral spring, one end being fastened (pin-jointed) at C and the other being attached to a central spindle. Take coordinate axes through C such that the x-axis passes through the spindle and let Q and P be the components of the reaction at C in the x and y directions respectively due to a couple M applied to the spindle.

Consider a small portion of the spring AB whose length is δL. The bending moment on this portion is $(Px - Qy)$ and, from the relationships established in Volume 1, Chapter 6, the change in slope (δi)

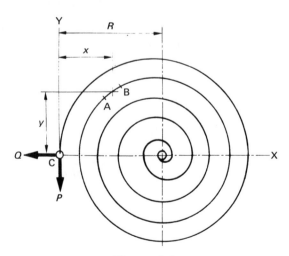

Figure 4.4

over the length AB is,

$$\delta i = \frac{\text{(bending moment)} \times \text{(length)}}{EI}$$

$$= \frac{1}{EI} \times (Px - Qy) \times \delta L$$

The total angle turned through, ϕ radians, is the summation of all such elements. Thus

$$\phi = \int \frac{1}{EI} \times (Px - Qy) \, dL$$

$$= \frac{1}{EI} \left(P \int x \, dL - Q \int y \, dL \right) \tag{i}$$

The integrals in (i) are to be taken over the whole length of the spring and are the first moments of this length about the y and x axes respectively. Thus

$$\int x \, dL = \bar{x} \times L \quad \text{and} \quad \int y \, dL = \bar{y} \times L$$

where $\bar{x}$ and $\bar{y}$ are the centroidal distances of the spring's profile from these axes. The centroid of the spring is approximately at the centre of the spindle and, assuming this to be *exactly* so, we have $\bar{y} = 0$ and $\bar{x} = R$ (the distance from C to the spindle). Thus from (i)

$$\phi = \frac{1}{EI} \times PR \times L$$

Taking moments about the spindle, M (the applied couple) $= PR$ and thus the angle turned through is

$$\phi = \frac{ML}{EI} \tag{ii}$$

The maximum bending moment occurs at D on the opposite side of the spindle to C (see Figure 4.4) and is approximately

$$M_{\text{max}} = 2R \times P = 2M$$

If the wire is rectangular in section, breadth b and thickness d the maximum stress is

$$\sigma_{\text{max}} = \frac{M_{\text{max}}}{Z} = \frac{2M}{\frac{1}{6}bd^2} = \frac{12M}{bd^2} \tag{iii}$$

From (iii) $M = bd^2 \sigma_{\text{max}}/12$ and using (ii) the strain energy is

$$U = \text{average couple} \times \text{angle turned through}$$

$$= \tfrac{1}{2}M\phi = \tfrac{1}{2}M \left(\frac{ML}{EI} \right) = \frac{M^2 L}{2EI} = \frac{1}{2} \left(\frac{bd^2 \sigma_{\text{max}}}{12} \right)^2 \times \frac{L}{E \times bd^3/12}$$

$$= \frac{\sigma_{\text{max}}^2}{24E} \times bdL = \frac{\sigma_{\text{max}}^2}{24E} \times \text{volume of the spring}$$

With the figures given in the question we have, from (iii),

$$M = \frac{bd^2\sigma_{max}}{12}$$

$$= \tfrac{1}{12}[(25 \times 10^{-3}\,\text{m}) \times (1 \times 10^{-3}\,\text{m})^2 \times (150 \times 10^6\,\text{N/m}^2)]$$

$$= 0.3125\,\text{N m}$$

For one complete turn of the spindle $\phi = 2\pi$ radians and, from (ii),

$$L = \frac{\phi EI}{M} = \frac{2\pi \times (200 \times 10^9\,\text{N/m}^2)}{0.3125\,\text{N m}} \times \left(\frac{25 \times 1^3}{12} \times 10^{-12}\,\text{m}^4\right)$$

$$= 8.38\,\text{m}$$

The strain energy can be calculated in two ways. Either

$$U = \frac{\sigma_{max}^2}{24E} \times \text{volume of the spring}$$

$$= \frac{(150 \times 10^6\,\text{N/m}^2)^2}{24 \times (200 \times 10^9\,\text{N/m}^2)} \times (25 \times 10^{-3}\,\text{m})$$

$$\times (1 \times 10^{-3}\,\text{m}) \times (8.38\,\text{m})$$

$$= 0.982\,\text{N m (or J)} \qquad\qquad (Ans)$$

or

$$U = \tfrac{1}{2}M\phi$$

$$= \tfrac{1}{2} \times 0.3125 \times 2\pi$$

$$= 0.982\,\text{N m (or J)} \qquad\qquad (Ans)$$

Problems

1 For springs that depend on the bending of a wire or plate, the average strain energy per unit volume can be expressed as σ^2/kE, where σ is the maximum bending stress. Determine the value of k for each of the following:

(a) Helical spring of round wire subjected to an axial couple;
(b) Semi- or quarter-elliptic leaf spring;
(c) Flat spiral spring of rectangular cross-section subjected to an axial couple.

Answer   (a) 8; (b) 6; (c) 24.

2 Assuming, without proof, the formulae for the strength and stiffness of a shaft subject to pure torsion, show that the torsional resilience of a shaft is equal to $\tau_{max}^2/4G$ times the volume of the shaft where τ_{max} is the maximum shear stress and G the modulus of rigidity. A spring with a small helix angle and 10 active coils is required to carry a load of 9 kN when compressed axially by 18 mm. Find the mean coil diameter and the diameter of spring section for a maximum shearing stress of 400 MN/m² ($G = 80 \times 10^9$ N/m²). (*IMechE*)

Answer   Coil diameter, 38.7 mm; wire diameter, 13.0 mm.

3 Derive a formula for the strain energy, in terms of the maximum bending stress, the volume of the spring and E, stored in a closely-coiled helical spring when subjected to an axial twisting moment.

A spring with 8 free coils of mean diameter 60 mm and 6 mm × 6 mm wire section, is used as a flexible coupling for a light direct drive between a motor and a machine. If the bending stress in the spring section is limited to 80 MN/m², find the greatest horse-power which may be transmitted at 1 500 rev/min. How much strain energy is stored in the spring under the above conditions? ($E = 200$ GN/m²). (*IMechE*)

Answer $U = \dfrac{\sigma_{max}^2}{8E} \times$ volume (for round wire) and $\dfrac{\sigma_{max}^2}{6E} \times$ volume

(for rectangular wire); 452 W; 0.290 N m.

4 The particulars of the close-coiled helical springs, shown in the assembly in Figure 4.5, are as follows:

Spring	Length uncompressed	Number of free coils	Mean coil diameter	Diameter of section
A	200 mm	12	100 mm	10 mm
B	175 mm	15	75 mm	6 mm

Find the pressure exerted by the disc C on the shoulder D. What force must be applied to the rod to hold the disc 12.5 mm from the shoulder? Also find the shear stresses, induced in the springs, when the disc is in this latter position. ($G = 80 \times 10^9$ N/m²). (*IMechE*)

Answer   348 N;   487 N; τ_{max} (in A) = 132.6 MN/m²; τ_{max} (in B) = 30.2 MN/m².

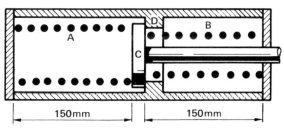

Figure 4.5

5 A laminated spring of the semi-elliptic type has a span of 675 mm and is built of leaves 8 mm thick by 45 mm wide. How many leaves would be required to carry a central load of 4.4 kN without the stress in the steel exceeding $220 \, \text{MN/m}^2$? What will be the deflection at the centre due to this load? Any formulae used should be derived. $E = 200 \times 10^9 \, \text{N/m}^2$. (*IMechE*)

Answer 8 leaves; 13.77 mm.

6 In the steel cantilever and spring arrangement, shown in Figure 4.6, an increasing load W is applied to deflect the cantilever and compress the spring. When W is zero there is a gap x between the top of the spring and the underside of the beam and when W is equal to 16 N the downward movement of the free end of the cantilever is 6 mm ($x < 6$ mm). If the close-coiled helical spring has 16 active coils, a mean coil diameter of 75 mm, and diameter of wire section 6 mm, find the value of x and the load on the cantilever when the gap is just closed. $E = 200 \times 10^9 \, \text{N/m}^2$; $G = 80 \times 10^9 \, \text{N/m}^2$. (*IMechE*)

Answer $x = 1.57$ mm; 1.97 N.

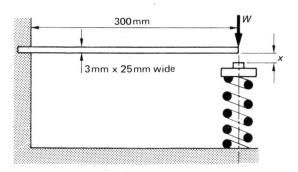

Figure 4.6

7 Prove that the energy stored per unit volume of a compressed helical spring made of round wire is $\tau^2/4G$, where τ is the maximum shearing stress in the wire and G is the modulus of rigidity.

Determine the mass of such a spring which requires a force of 9 kN to produce a compression of 140 mm, the maximum shearing stress being $450 \, \text{MN/m}^2$.

Density of material, $7.83 \, \text{Mg/m}^3$.

Modulus of rigidity, $82 \times 10^9 \, \text{N/m}^2$. (*UL*)

Answer 7.99 kg.

8 Wire of circular section (dia. = 8 mm) is available to make a close-coiled helical spring of a stiffness of 20 kN/m. The spring will not be required to carry more than 280 N static load, and the maximum allowable stress in the wire is $80 \, \text{MN/m}^2$. If the modulus of rigidity equals $80 \, \text{GN/m}^2$, determine the mean radius of the coils and the number of coils required. (*UL*)

Answer 28.7 mm; 10.8 coils.

9 A close-coiled helical steel spring is required having a sliding fit over a rod of 28 mm diameter. The spring is to carry a maximum axial load of 120 N and the deflection at this load is to be 20 mm; the shearing stress must not exceed 200 MN/m².

Determine the diameter of the wire required, the mean diameter of the coil and the number of turns necessary. Take $G = 80$ GN/m². *(UL)*

Answer 3.15 mm (root of cubic equation); 31.15 mm; 5.43 coils.

10 Make a sketch of a leaf spring showing the shape to which the ends of the plates should be made and give the reason for doing this.

A leaf spring which carries a central load of 9 kN consists of plates each 76 mm wide and 8 mm thick. If the length of the spring is 900 mm, determine the least number of plates required if the maximum stress due to bending is limited to 220 MN/m² and the maximum deflection must not exceed 32 mm. Take $E = 200$ GN/m².

Find for the number of plates obtained, the actual values of the maximum stress and maximum deflection and also the radius to which the plates should be formed if they are to straighten under the given load. *(UL)*

Answer 12 plates; 208.2 MN/m²; 26.3 mm; 3.84 m.

11 A laminated steel spring, simply supported at the ends and centrally loaded, with a span of 750 mm, is required to carry a proof load of 8 kN, and the central deflection is not to exceed 50 mm. The bending stress must not be greater than 400 MN/m². Plates are available in multiples of 1 mm for thickness and in multiples of 4 mm for width.

Determine suitable values for the thickness, width and number of plates and the radius to which the plates should be formed. Assume the width to be ten times the thickness. ($E = 200$ GN/m².) *(UL)*

Answer 11 plates, each 6 mm × 60 mm; initial radius 1.584 m.

12 A close-coiled helical spring is to have a stiffness of 8.75 N per cm of compression, a maximum load of 40 N, and a maximum shearing stress of 125 MN/m². The solid length of the spring (i.e. when the coils are touching) is to be 46 mm. Find the diameter of the wire, the mean radius of the coils and the number of coils required.

Modulus of rigidity = 40 GN/m². *(UL)*

Answer 2.70 mm; 12.13 mm; 17.

13 A close-coiled helical spring of circular section extends unit length when subjected to an axial load W, and there is an angular rotation of 1 radian when a torque T is independently applied, about the axis of the spring. If Poisson's ratio is v and the mean diameter of the coil is D, show that

$$T/W = \tfrac{1}{4}D^2(1 + v)$$

Determine Poisson's ratio for the material of a spring of 75 mm mean diameter of coils, if 240 N extends the spring 130 mm and a torque of 3.4 N m produces an angular rotation of 60°.   (*UL*)
Answer $v = 0.25$.

14   A composite spring has two close-coiled helical steel springs in series; each spring has a mean coil diameter of 8 times the diameter of its wire. One spring has 20 coils and wire diameter 2.5 mm. Find the diameter of the wire in the other if it has 15 coils and the stiffness of the composite spring is 1.25 kN/m.

Find the greatest axial load that can be applied to the spring and the corresponding extension for a maximum shearing stress of 300 MN/m². ($G = 80 \times 10^9$ N/m².)   (*UL*)
Answer $d = 1.96$ mm; 56.6 N; 45.3 mm.

15   A carriage spring, centrally loaded and simply supported at the ends, has 10 steel plates, each 50 mm wide by 6 mm thick. If the longest plate is 750 mm long, find the initial radius of curvature of the plates when the greatest bending stress is 150 MN/m² and the plates are finally straight. Neglecting the loss of energy at impact, determine the greatest height from which a mass weighing 200 N may be dropped centrally on the spring without exceeding the limiting bending stress of 150 MN/m². ($E = 200$ GN/m².)   (*UL*)
Answer   4 m; 87.9 mm.

16   A quarter-elliptic, i.e. cantilever, leaf spring has a length of 500 mm and consists of plates each 50 mm wide and 6 mm thick. Find the least number of plates which can be used if the deflection under a gradually applied load of 2 kN is not to exceed 70 mm.

If instead of being gradually applied, the load of 2 kN falls a distance of 6 mm on to the undeflected spring, find the maximum deflection and stress produced. ($E = 200$ GN/m².)   (*UL*)
Answer   10 plates; 144.6 mm; 694 MN/m².

17   Derive the expression for the strain energy of a carriage spring in terms of the maximum bending stress, the volume of the spring and the elastic modulus E. The spring is supported at the ends and loaded at the centre, and the ends of the plates are pointed in the usual manner.

A carriage spring, as above, is to be constructed using plates 60 mm wide and 6 mm thick and is to carry a suddenly applied central load of 4 kN. The maximum values of stress and deflection produced by this load are not to exceed 420 MN/m² and 50 mm respectively. Determine the length of the spring and the number of plates required. $E = 200$ GN/m².   (*UL*)

Answer $\dfrac{\sigma_{\max}^2}{2E} \times$ volume; 756 mm; 10 plates.

18 A flat spiral spring is made of steel 12 mm broad and 0.5 mm thick. The end at the greatest radius is attached to a fixed point and the other end to a spindle. The length of the steel strip is 6 m. Determine:

(a) the maximum turning moment which can be applied to the spindle if the stress in the strip is not to exceed $550 \, MN/m^2$;

(b) the number of turns required to be given to the spindle;
(c) the energy then stored in the spring. ($E = 200 \, GN/m^2$).

(*UL*)

Answer (a) 0.137 5 N m. (b) 5.25 turns. (c) 2.27 N m (or J).

19 An instrument control spring is made of phosphor-bronze 1 mm wide and 0.1 mm radial thickness. It is to be formed into a flat spiral spring pinned at the outer end and at the inner end to the collet on the instrument arbor.

Calculate the necessary length of spring so that a torque of 450 dyne-cm will cause a rotation of 90 degrees.

E for phosphor-bronze $= 113 \, GN/m^2$.

(Note that 1 newton $= 10^5$ dynes.) (*UL*)

Answer 329 mm.

20 A flat spiral spring is formed from a strip of material of rectangular section 12 mm wide and 0.8 mm thick and of total length 3.6 m. It is pinned at its outer end and a winding couple of 0.25 N m is applied at the central spindle. Find (a) the maximum stress and (b) the strain energy stored in the material.

Work from first principles. Take $E = 200 \, GN/m^2$. (*UL*)

Answer (a) $391 \, MN/m^2$; (b) 1.10 N m.

5

Thick cylinders and rotating discs

Principal stresses in thick-walled cylinder

At a point in the wall of a cylinder subjected to internal (or external) pressure the principal stresses are:

(a) a longitudinal stress σ_L
(b) a hoop or circumferential stress σ_H
(c) a radial stress σ_R

Sign convention

In this chapter tensile stresses are taken as positive. The radial stress σ_R is invariably compressive and therefore negative.

Longitudinal stress and strain

The strain in the direction of σ_L is

$$\frac{\sigma_L}{E} - \frac{v}{E}(\sigma_H + \sigma_R)$$

It is assumed that plane cross-sections remain plane whether or not the longitudinal tension is borne by the cylinder. It follows that the above strain is the same at all radii and, in a given example, σ_L is constant and $(\sigma_H + \sigma_R)$ is constant.

Variation of hoop and radial stresses

At radius r the hoop and radial stresses are:

$$\sigma_H = A + \frac{B}{r^2} \quad \text{and} \quad \sigma_R = A - \frac{B}{r^2}$$

where A and B are constants to be determined from the conditions at the inner and outer surfaces. In particular, at the inner surface of a cylinder subjected to an internal pressure P, $\sigma_R = -P$ and, at the outer surface, $\sigma_R = 0$.

Cylinder with internal pressure P

The hoop and radial stresses at radius r are given by

$$\sigma_H = \left\{\frac{(R_2/r)^2 + 1}{(R_2/R_1)^2 - 1}\right\}P \qquad \sigma_R = -\left\{\frac{(R_2/r)^2 - 1}{(R_2/R_1)^2 - 1}\right\}P$$

where R_1 and R_2 are the internal and external radii respectively.

Cylinder with external pressure P

The hoop and radial stresses at radius r are given by

$$\sigma_H = -\left\{\frac{1 + (R_1/r)^2}{1 - (R_1/R_2)^2}\right\}P \qquad \sigma_R = -\left\{\frac{1 - (R_1/r)^2}{1 - (R_1/R_2)^2}\right\}P$$

Maximum hoop stress

The maximum hoop stress occurs at the inner surface. For internal pressure P and zero external pressure

$$\sigma_{H(\text{max})} = \left\{\frac{(R_2/R_1)^2 + 1}{(R_2/R_1)^2 - 1}\right\}P = \left(\frac{k^2 + 1}{k^2 - 1}\right)P$$

where $k = R_2/R_1$.

For external pressure P and zero internal pressure

$$\sigma_{H(\text{max})} = -\left\{\frac{2}{1 - (R_1/R_2)^2}\right\}P = -\left(\frac{2}{1 - 1/k^2}\right)P$$

Solid shaft with external pressure

At all points in a solid shaft subjected to an external pressure P,

$$\sigma_H = \sigma_R = -P$$

Rotating disc

At radius r in a thin circular disc of uniform density ρ rotating at angular velocity ω about its axis, the hoop and radial stresses are

$$\sigma_H = A + \frac{B}{r^2} - \tfrac{1}{8}\rho r^2 \omega^2 (1 + 3v)$$

$$\sigma_R = A - \frac{B}{r^2} - \tfrac{1}{8}\rho r^2 \omega^2 (3 + v)$$

where $v =$ Poisson's ratio.

Worked examples

5.1 Stress and diameter change at surfaces of thick cylinder with internal pressure

Two strain gauges are attached to the outer surface of a cylinder so as to measure the longitudinal and circumferential strains. The cylinder has closed ends and its internal and external diameters are 150 mm and 200 mm respectively. Under a certain internal pressure the gauge readings indicate that the longitudinal and circumferential stresses are 36 MN/m^2 and 72 MN/m^2 both tensile. Find:

   (a) the internal pressure,
   (b) the circumferential or hoop stress at the inner surface,
   (c) the change in the internal diameter due to the pressure.

Take $E = 206$ MN/m^2 and $v = 0.28$.

Solution (a) On the assumption that cross-sections remain plane, the longitudinal stress σ_L is uniform through the wall of the cylinder. Let P = internal pressure. Then, for equilibrium,

force on end due to P = force in wall due to σ_L

$$P \times \frac{\pi}{4} \times 0.15^2 = \sigma_L \times \frac{\pi}{4}(0.2^2 - 0.15^2)$$

and

$$P = \frac{36\,\text{MN/m}^2 \times (0.2^2 - 0.15^2)}{0.15^2}$$

$$= 28\,\text{MN/m}^2 \quad \text{or} \quad 280\,\text{bar} \qquad (Ans)$$

since $1\,\text{bar} = 10^5\,\text{N/m}^2$.

(b) A further consequence of the assumption that cross-sections remain plane is that $(\sigma_H + \sigma_R)$ is constant through the wall. At the outer surface the pressure is zero and hence $\sigma_R = 0$. Furthermore, $\sigma_H = 72\,\text{MN/m}^2$. The value of $(\sigma_H + \sigma_R)$ is therefore $(72 + 0) = 72$ and this holds for all radii. At the inner surface, $\sigma_R = -$(internal pressure) $= -28\,\text{MN/m}^2$ and:

$$\sigma_H - 28 = 72$$

so that

$$\sigma_H = 100\,\text{MN/m}^2 \qquad (Ans)$$

(c) At the inner surface, using the results obtained above,

$$\text{circumferential strain} = \frac{\sigma_H}{E} - \frac{v\sigma_R}{E} - \frac{v\sigma_L}{E}$$

$$= \frac{100 \times 10^6 - 0.28(-28 \times 10^6 + 36 \times 10^6)}{206 \times 10^9}$$

$$= 0.000\,474$$

Since the proportional change in diameter equals the proportional change in circumference, we have:

$$\text{change in diameter} = 0.000\,474 \times 150$$

$$= 0.071\,2\,\text{mm (increase)} \qquad (Ans)$$

5.2 Lamé formulae for hoop and radial stresses in thick cylinder

Develop the Lamé relationships for the radial and circumferential stresses in a thick-walled cylinder due to internal or external pressure.

A cylinder of 200 mm internal diameter is to withstand an internal pressure of $40\,\text{MN/m}^2$. Determine the necessary external diameter if the maximum hoop stress in the wall is to be $100\,\text{MN/m}^2$.

Solution Let σ_R be the radial stress and σ_H the circumferential stress at radius r within the cylinder wall. The following theory (Lamé's) leads to generalized equations relating σ_R, σ_H and r, and involving numerical constants whose values, in particular examples, are determined from boundary conditions such as the pressures on the inner and outer surfaces. The radial stress is always compressive as shown in Figure 5.1(a) and the equations can be deduced from the equilibrium of a small element ABCD shown in that diagram. The work is simplified, however, by considering half of an annular ring of the cylinder having internal radius r and thickness δr as shown in Figure 5.1(b). In this diagram all stresses are taken as positive if tensile and the radial stress σ_R can be regarded as a negative internal pressure on the ring. Similarly there is a negative external pressure $\sigma_R + \delta\sigma_R$ at radius $r + \delta r$.

The conditions for the equilibrium of this half-ring are similar to those used in thin cylinder theory (see Volume 1, Chapter 10).

For unit axial length, the total downward force (Figure 5.1(b)) is

$$2r\sigma_R + 2\sigma_H\,\delta r$$

and the total upward force is

$$2(r + \delta r)(\sigma_R + \delta\sigma_R)$$

Equating these two quantities

$$2(r + \delta r)(\sigma_R + \delta\sigma_R) = 2r\sigma_R + 2\sigma_H\,\delta r$$
$$r\sigma_R + \sigma_R\delta r + r\delta\sigma_R + \delta r\delta\sigma_R = r\sigma_R + \delta_H\delta r$$
$$r\delta\sigma_R + \sigma_R\delta r - \sigma_H\delta r = 0$$

neglecting products of small quantities. Dividing through by δr and proceeding to the limit

$$r\frac{d\sigma_R}{dr} + \sigma_R - \sigma_H = 0 \qquad\qquad\text{(i)}$$

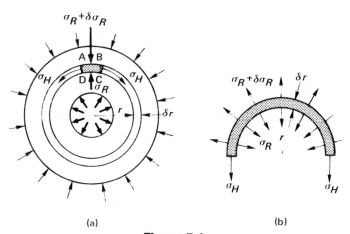

(a)   (b)

Figure 5.1

A second relationship between σ_R and σ_H derives from the assumption of uniform longitudinal strain. As stated at the head of this chapter it follows that $\sigma_R + \sigma_H$ is a constant and it is convenient in the working which follows to denote this constant by $2A$.

Thus

$$\sigma_R + \sigma_H = 2A \qquad \text{(ii)}$$

Substituting the value of σ_H from (ii) in (i) we obtain

$$r\frac{d\sigma_R}{dr} + \sigma_R - (2A - \sigma_R) = 0$$

$$r\frac{d\sigma_R}{dr} = 2(A - \sigma_R)$$

Separating differentials

$$\frac{d\sigma_R}{A - \sigma_R} = 2\frac{dr}{r}$$

Integrating both sides

$$-\ln(A - \sigma_R) = 2\ln r + \text{constant}$$

$$= \ln r^2 + \text{constant}$$

$$-\ln[r^2(A - \sigma_R)] = \text{constant}$$

$$r^2(A - \sigma_R) = B \quad \text{(a different constant)}$$

Rearranging,

$$\sigma_R = A - \frac{B}{r^2} \qquad \text{(iii)}$$

and, using (ii),

$$\sigma_H = 2A - \sigma_R = 2A - \left(A - \frac{B}{r^2}\right) = A + \frac{B}{r^2} \qquad \text{(iv)}$$

The constants A and B are found, in a given example, by using known boundary conditions, e.g. the radial stress is taken as zero at a surface exposed to atmospheric conditions.

For the numerical part of the question the circumferential stress is greatest at the inner surface. The radial stress at the same surface is numerically equal to the internal pressure but is taken as negative in accordance with the sign convention adopted above. Thus, working in MN/m^2 for stress and metres for radius, we have:

when $r = 0.1$, $\sigma_H = 100$ and $\sigma_R = -40$

Substituting these values in the stress relationships (iii) and (iv),

$$100 = A + \frac{B}{0.1^2}$$

and

$$-40 = A - \frac{B}{0.1^2}$$

Solving these equations, $A = 30$ and $B = 0.7$.
At the outer surface, $\sigma_R = 0$ and hence

$$0 = 30 - \frac{0.7}{r^2}$$

from which $r = 0.153$ m
The required external diameter is therefore 306 mm.   (*Ans*)

5.3 Comparison of thick and thin cylinder theories for maximum hoop stress in thick cylinder due to internal pressure

Obtain an expression for the maximum hoop stress for a cylinder subjected to an internal pressure P in terms of k the ratio of the external to the internal diameter.
    Find the value of k if the maximum hoop stress is to be

(a) $3 \times$ internal pressure;
(b) $20 \times$ internal pressure.

Also solve case (b) using thin cylinder theory.

Solution Let R_1 and R_2 be the internal and external radii respectively. The radial and hoop stresses at radius r are, as shown in Example 5.2,

$$\sigma_R = A - \frac{B}{r^2} \quad \text{and} \quad \sigma_H = A + \frac{B}{r^2}$$

At the inside surface $\sigma_R = -P$ (the internal pressure). Thus, putting $r = R_1$,

$$-P = A - \frac{B}{R_1^2} \tag{i}$$

Also, at the outside surface the radial stress is zero (atmospheric pressure) and, putting $r = R_2$,

$$0 = A - \frac{B}{R_2^2} \tag{ii}$$

Subtracting (i) from (ii),

$$P = \frac{B}{R_1^2} - \frac{B}{R_2^2} = B\left(\frac{R_2^2 - R_1^2}{R_1^2 R_2^2}\right)$$

$$B = P\left(\frac{R_1^2 R_2^2}{R_2^2 - R_1^2}\right) \tag{iii}$$

Substituting this value of B in (ii), we obtain

$$A = \frac{B}{R_2^{\,2}} = P\left(\frac{R_1^{\,2}}{R_2^{\,2} - R_1^{\,2}}\right) \tag{iv}$$

Using (iii) and (iv) the hoop stress at any radius r is

$$\sigma_H = A + \frac{B}{r^2} = P\left(\frac{R_1^{\,2}}{R_2^{\,2} - R_1^{\,2}}\right) + \frac{P}{r^2}\left(\frac{R_1^{\,2}R_2^{\,2}}{R_2^{\,2} - R_1^{\,2}}\right)$$

$$= \frac{PR_1^{\,2}}{R_2^{\,2} - R_1^{\,2}}\left(\frac{R_2^{\,2}}{r^2} + 1\right) \tag{v}$$

The maximum value of σ_H clearly corresponds to the smallest value of r, i.e. R_1 (at the inside surface). Hence, the maximum hoop stress is

$$\sigma_{H(max)} = \frac{PR_1^{\,2}}{R_2^{\,2} - R_1^{\,2}}\left(\frac{R_2^{\,2}}{R_1^{\,2}} + 1\right)$$

$$= \frac{PR_1^{\,2}}{R_2^{\,2} - R_1^{\,2}}\left(\frac{R_2^{\,2} + R_1^{\,2}}{R_1^{\,2}}\right)$$

$$= P\left(\frac{R_2^{\,2} + R_1^{\,2}}{R_2^{\,2} - R_1^{\,2}}\right)$$

$$= P\left(\frac{k^2 + 1}{k^2 - 1}\right) \tag{vi}$$

since the ratio of the radii equals the ratio of the diameters.

(a) If $\sigma_{H(max)} = 3P$, then

$$3 = \left(\frac{k^2 + 1}{k^2 - 1}\right)$$

$$3(k^2 - 1) = k^2 + 1 \qquad 2k^2 = 4$$

$$k = \sqrt{2} = 1.414 \tag{Ans}$$

(b) If $\sigma_{H(max)} = 20P$,

$$20 = \left(\frac{k^2 + 1}{k^2 - 1}\right)$$

$$19k^2 = 21$$

$$k = \sqrt{(21/19)} = 1.051 \tag{Ans}$$

(c) Using thin cylinder theory to solve (b), the hoop stress is $\sigma_H = Pd/2t$ and, taking d as the internal diameter, $\sigma_H = 20P$. Thus

$$20 = d/2t \qquad d/t = 40$$

$$k = \frac{\text{external diameter}}{\text{internal diameter}} = \frac{d + 2t}{d}$$

$$= 1 + \frac{2t}{d} = 1 + \frac{2}{40}$$

$$= 1.05 \tag{Ans}$$

5.4 Radial pressure between steel sleeve and shaft

A steel sleeve of length 12 cm and 12 cm external diameter is shrunk on to a steel shaft of diameter 8 cm such that the maximum tensile hoop stress induced in the sleeve is 130 MN/m². Determine the resulting normal pressure, in MN/m², at the mating surface. Plot graphs showing the distributions of the radial and hoop stresses throughout the thickness of the sleeve.

(*IMechE*)

Solution In the notation of Example 5.3, $k = 3/2$. Hence using equation (vi) of that example

$$130 \text{ MN/m}^2 = P\left[\frac{(3/2)^2 + 1}{(3/2)^2 - 1}\right]$$

and the normal pressure at the mating surface is

$$P = 130 \text{ MN/m}^2 \times \left[\frac{(3/2)^2 - 1}{(3/2)^2 + 1}\right]$$

$$= 50 \text{ MN/m}^2 \qquad\qquad\qquad (Ans)$$

Using the Lamé formulae

$$\sigma_R = A - \frac{B}{r^2} \quad \text{and} \quad \sigma_H = A + \frac{B}{r^2}$$

we have, working in MN/m² for stress and cm for radius,

$$\sigma_H = 130 \quad \text{when} \quad r = 4 \quad \text{or} \quad 130 = A + \frac{B}{4^2} \qquad (i)$$

Also

$$\sigma_R = 0 \quad \text{when} \quad r = 6 \quad \text{or} \quad 0 = A - \frac{B}{6^2} \qquad (ii)$$

Solving the simultaneous equations (i) and (ii),

$$A = 40 \quad \text{and} \quad B = 1\,440$$

Using these values the radial and hoop stresses are given by

$$\sigma_R = 40 - \frac{1\,440}{r^2} \quad \text{and} \quad \sigma_H = 40 + \frac{1\,440}{r^2}$$

The results for σ_R are negative, indicating compressive stress and, tabulating values,

r cm	σ_R MN/m² (compressive)	σ_H MN/m² (tensile)
4	50	130
$4\frac{1}{2}$	31	111
5	17.6	97.6
$5\frac{1}{2}$	7.5	87.5
6	0	80

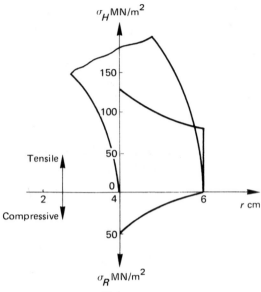

σ_H MN/m^2

150

100

50

Tensile

0

2   4   6

r cm

Compressive

50

σ_R MN/m^2

Figure 5.2

Figure 5.2 shows the distribution of the radial and hoop stresses throughout the thickness of the sleeve.

5.5 Stresses due to shrinkage and internal pressure in compound tube

A compound tube is made by shrinking one tube on another. The common diameter is 6 cm, the shrinkage allowance (based on diameter) is 0.01 mm and each tube is 1 cm thick. If both tubes are of steel $(E = 200 \text{ GN/m}^2)$ calculate the radial pressure between the tubes.

Draw a diagram to show the distribution of the hoop stress throughout the wall:

(a) due to the shrinkage, and
(b) due to an internal pressure of 60 MN/m^2

Show also the resultant hoop stress distribution.

Solution Working in MN/m^2 for stress and cm for radius throughout, let P be the radial pressure between the tubes due to shrinkage.

For the outer tube, the diameter ratio $k = \frac{4}{3}$ and the hoop stress at the common radius is

$$P\left(\frac{k^2 + 1}{k^2 - 1}\right) = \frac{25P}{7} \ell \tag{i}$$

For the inner tube, take

$$\sigma_R = A - \frac{B}{r^2} \quad \text{and} \quad \sigma_H = A + \frac{B}{r^2}$$

When $r = 3$, $\sigma_R = -P$, and

$$-P = A - \frac{B}{3^2} \qquad \text{(ii)}$$

Also, when $r = 2$, $\sigma_R = 0$, and

$$0 = A - \frac{B}{2^2} \qquad \text{(iii)}$$

Subtracting (ii) from (iii),

$$P = \frac{B}{9} - \frac{B}{4} = -\frac{5B}{36}$$

Hence

$$B = -36P/5$$

and, substituting in (iii),

$$A = -9P/5$$

Using these values of A and B, the hoop stress in the inner tube at the common radius ($r = 3$) is

$$-\frac{9P}{5} - \frac{36P}{5 \times 3^2} = -\frac{13P}{5} \qquad \text{(iv)}$$

Thus, at the common radius, tensile hoop strain in the outer tube is

$$\frac{\sigma_H}{E} + \frac{vP}{E} = \frac{25P}{7E} + \frac{vP}{E}$$

Tensile hoop strain in inner tube is

$$-\frac{13P}{5E} + \frac{vP}{E}$$

Total strain difference

$$= \text{(tensile hoop strain in outer tube)}$$
$$\quad - \text{(tensile hoop strain in inner tube)}$$
$$= \left(\frac{25P}{7E} + \frac{vP}{E}\right) - \left(-\frac{13P}{5E} + \frac{vP}{E}\right) = \frac{P}{E}\left(\frac{25}{7} + \frac{13}{5}\right) = \frac{216P}{35E}$$

This strain difference equals the ratio of the shrinkage allowance to the common diameter, since the proportional change in diameter equals the proportional change in circumference. Thus

$$\frac{216P}{35E} = \frac{0.01 \times 10^{-1}}{6}$$

$$P = \frac{0.001}{6} \times \frac{35 \times 200 \times 10^9}{216} = 5.4 \times 10^6 \, \text{N/m}^2$$

The radial pressure between the tubes is $5.4\,\text{MN/m}^2$. \hfill (*Ans*)

Using this value of P the constants for the inner tube are

$$A = -9P/5 = -9 \times 5.4/5 = -9.72$$
$$B = -36P/5 = -36 \times 5.4/5 = -38.88$$

With these values, the hoop stress at any radius r in the inner tube is

$$\sigma_H = A + \frac{B}{r^2} = -9.72 - \frac{38.88}{r^2} \tag{v}$$

Since these values of the constants A and B apply to the inner tube only, we must for the outer tube put

$$\sigma_R = A' - \frac{B'}{r^2} \qquad \sigma_H = A' + \frac{B'}{r^2}$$

When $r = 3$, $\sigma_R = -P = -5.4$, and thus

$$-5.4 = A' - \frac{B'}{3^2} \tag{vi}$$

When $r = 4$, $\sigma_R = 0$, and thus

$$0 = A' - \frac{B'}{4^2} \tag{vii}$$

Solving the simultaneous equations (vi) and (vii)

$$A' = 6.944 \quad \text{and} \quad B' = 111.1$$

With these values the hoop stress at any radius in the outer tube is

$$\sigma_H = 6.944 + \frac{111.1}{r^2} \tag{viii}$$

The stresses due to the internal pressure can be calculated by regarding the two tubes as a single cylinder (provided they are of the same material) with internal and external diameters 4 cm and 8 cm respectively. Hence, due to the internal pressure

$$\sigma_R = A'' - \frac{B''}{r^2} \quad \text{and} \quad \sigma_H = A'' + \frac{B''}{r^2}$$

throughout both tubes.

When $r = 2$, $\sigma_R = -60$. Thus

$$-60 = A'' - \frac{B''}{2^2} \tag{ix}$$

Also, when $r = 4$, $\sigma_R = 0$. Thus

$$0 = A'' - \frac{B''}{4^2} \tag{x}$$

Solving the simultaneous equations (ix) and (x)

$$A'' = 20 \quad \text{and} \quad B'' = 320$$

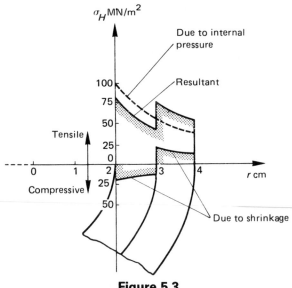

σ_H MN/m^2

Due to internal pressure

Resultant

Due to shrinkage

Tensile

Compressive

r cm

Figure 5.3

and the hoop stress at any radius r is

$$\sigma_H = 20 + \frac{320}{r^2} \tag{xi}$$

Using equations (v), (viii) and (xi) the hoop stresses in MN/m^2 are:

r cm	Inner tube			Outer tube		
	2	$2\frac{1}{2}$	3	3	$3\frac{1}{2}$	4
Due to shrinkage Equation (v) Equation (viii)	-19.44	-15.94	-14.04	19.29	16.01	13.89
Due to pressure Equation (xi)	100	71.2	55.56	55.56	46.12	40
Resultant stress	80.56	55.26	41.52	74.85	62.13	53.89

Figure 5.3 shows the distribution of the hoop stress due to the shrinkage and the internal pressure as well as the resultant stress.

(Note that the maximum resultant stress in the compound tube is 80.56 MN/m^2 compared with 100 MN/m^2 for a single tube of the same overall dimensions and internal pressure.)

5.6 Driving allowance for cast-iron hub on steel shaft

Determine the driving allowance for a cast-iron hub, 6 cm external diameter, on a steel shaft 4 cm diameter, if the maximum circumferential stress in the hub is to be 60 MN/m^2.
E for steel = 200 GN/m^2. E for cast-iron = 100 GN/m^2.
Poisson's ratio = 0.3 for both.

Solution The diameter ratio for the hub is $k = 6/4 = 3/2$. If $\sigma_{H(\text{max})}$ is the maximum circumferential stress in the hub and P is the radial pressure at the common surface, then

$$\sigma_{H(\text{max})} = P\left(\frac{k^2 + 1}{k^2 - 1}\right)$$

$$P = \sigma_{H(\text{max})}\left(\frac{k^2 - 1}{k^2 + 1}\right) = 60 \text{ MN/m}^2\left[\frac{(3/2)^2 - 1}{(3/2)^2 + 1}\right]$$

$$= 23.08 \text{ MN/m}^2$$

The formulae

$$\sigma_R = A - \frac{B}{r^2} \quad \text{and} \quad \sigma_H = A + \frac{B}{r^2}$$

are valid for a shaft subjected to external pressure since the shaft can be regarded as a cylinder of zero internal radius. However the formulae must apply when $r = 0$ and, since the stress is not infinite at the centre, $B = 0$.

At all points in the shaft, therefore $\sigma_R = \sigma_H = A$. At the outside surface $\sigma_R = -P = -23.08 \text{ MN/m}^2$.

Hence, working in MN/m^2 units, at all points in the shaft

$$\sigma_R = \sigma_H = -23.08$$

At the common diameter, if suffixes s and c refer to steel and cast iron respectively, we have

Tensile hoop strain in hub:

$$\frac{(\sigma_H)_c}{E_c} - \frac{v(\sigma_R)_c}{E_c}$$

$$= \frac{60}{100 \times 10^3} - \frac{0.3 \times (-23.08)}{100 \times 10^3} = \frac{66.92}{10^5}$$

Tensile hoop strain in shaft:

$$\frac{(\sigma_H)_s}{E_s} - \frac{v(\sigma_R)_s}{E_s}$$

$$= \frac{-23.08}{200 \times 10^3} - \frac{0.3 \times (-23.08)}{200 \times 10^3} = -\frac{8.08}{10^5}$$

Total strain difference

$$= (\text{hoop strain in hub}) - (\text{hoop strain in shaft})$$

$$= \frac{66.92}{10^5} - \frac{-8.08}{10^5} = 75 \times 10^{-5}$$

Required driving allowance

$$= \text{total strain difference} \times \text{diameter}$$

$$= (75 \times 10^{-5}) \times 4 \text{ cm}$$

$$= 0.003 \text{ cm} \quad \text{or} \quad 0.03 \text{ mm} \qquad \qquad (\textit{Ans})$$

5.7 Axial load on shaft and sleeve assembly

A steel sleeve is pressed on to a solid steel shaft of 5 cm diameter. The radial pressure between the shaft and the sleeve is 20 MN/m² and the hoop stress at the inner surface of the sleeve is 48 MN/m². If an axial compressive load of 60 kN is now applied to the shaft determine the alteration of radial pressure.
Poisson's ratio = 0.304. *(UL)*

Solution Let P be the increase in radial pressure due to the compressive load.

Since the hoop stress at the inner surface of the sleeve bears a constant ratio to the radial pressure P, the increase in the hoop stress due to P is $(48/20) \times P = 2.4P$.

Also, for the shaft, the changes in the radial and hoop stresses due to P are given by $\sigma_R = \sigma_H = -P$.

At the common radius, due to P,

Increase in tensile hoop strain of the sleeve is

$$\left(\frac{\sigma_H}{E} - \frac{\nu \sigma_R}{E} \right) \text{ for sleeve}$$

$$= \frac{2.4P}{E} + \frac{0.304P}{E}$$

$$= 2.704P/E \tag{i}$$

and the increase in the hoop strain of the shaft is

$$\left(\frac{\sigma_H}{E} - \frac{\nu \sigma_R}{E} \right) \text{ for shaft}$$

$$= \left(\frac{-P}{E} + \frac{0.304P}{E} \right)$$

$$= \frac{-0.696P}{E} \tag{ii}$$

Also, the proportional increase in the diameter of the shaft due to the compressive load is

$$\nu \times \text{longitudinal strain} = 0.304 \times \text{stress}/E$$

$$= \frac{0.304 \times (60 \times 10^3 \, \text{N})}{\frac{1}{4}\pi \times (5 \times 10^{-2} \, \text{m})^2 \times E} \tag{iii}$$

Since the change in internal diameter of the sleeve must equal the change in diameter of the shaft, then

tensile hoop strain of sleeve due to P

= proportional increase in shaft diameter due to load
   + hoop strain in the shaft

or, from (i), (ii) and (iii),

$$\frac{2.704P}{E} = \frac{0.304 \times (60 \times 10^3 \, \text{N})}{\frac{1}{4}\pi(5 \times 10^{-2} \, \text{m})^2 \times E} - \frac{0.696 \, P}{E}$$

$$2.704P = 9.292 \times 10^6 - 0.696P$$

$$P = 2.73 \times 10^6 \, \text{N/m}^2$$

The alteration in radial pressure is 2.73 MN/m² (increase)　　　(*Ans*)

5.8　Radial stress in thin rotating disc

A thin rotating disc of uniform density ρ rotates at angular velocity ω. Show that the radial stress at any radius r is given by

$$\sigma_R = A - \frac{B}{r^2} - \tfrac{1}{8}(3 + v)\rho\omega^2 r^2$$

where v is Poisson's ratio and A and B are constants.

If the disc has an inner radius of $7\frac{1}{2}$ cm and an outer radius of 30 cm find the maximum radial stress at a speed of 3 600 rev/min. Take $E = 200 \, \text{GN/m}^2$, $v = 0.33$ and $\rho = 7.5 \, \text{Mg/m}^3$. 　　(*UL*)

Solution　Suppose (Figure 5.4) ABCD is a small element of a rotating disc at radius r, having radial width δr, and subtending a small angle θ at the centre O. Let σ_R and $(\sigma_R + \delta\sigma_R)$ be the radial stresses at radii r and $(r + \delta r)$, and σ_H the hoop or circumferential stress at radius r, tensile being positive in all cases. For unit thickness of material the forces on this element are as shown in the diagram. These forces are not in equilibrium and resolving towards O,

$$\text{Resultant force} = \sigma_R r\theta - (\sigma_R + \delta\sigma_R)(r + \delta r)\theta$$

$$+ 2\sigma_H \delta r \cos\left(90° - \tfrac{1}{2}\theta\right)$$

$$= -\sigma_R \delta r\theta - \delta\sigma_R r\theta + \sigma_H \delta r\theta$$

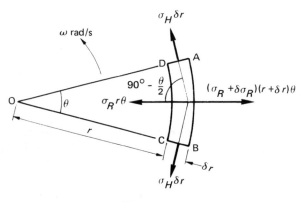

Figure 5.4

ignoring products of small quantities and noting that $\sin \frac{1}{2}\theta = \frac{1}{2}\theta$ for small angles.

The mass of the element is $\rho \times$ volume $= \rho r\theta \cdot \delta r$ and due to the rotational speed ω it has an acceleration $\omega^2 r$ towards O. Using the relationship, force $=$ mass $\times$ acceleration,

$$-\sigma_R \delta r\theta - \delta\sigma_R r\theta + \sigma_H \delta r\theta = (\rho r\theta\delta r) \times \omega^2 r$$

$$-\sigma_R - \frac{\delta\sigma_R}{\delta r}r + \sigma_H = \rho\omega^2 r^2$$

and, in the limit,

$$\sigma_H - \sigma_R = \rho\omega^2 r^2 + r\frac{d\sigma_R}{dr} \qquad\qquad \text{(i)}$$

A second relationship between σ_H and σ_R is obtained from considerations of strain. If a point at radius r in the unstressed disc moves radially outwards by an amount u the circumferential strain is

$$\frac{\text{increase in circumference}}{\text{original circumference}} = \frac{2\pi(r+u) - 2\pi r}{2\pi r} = \frac{u}{r}$$

and relating this to the stresses we have

$$\frac{u}{r} = \frac{\sigma_H}{E} - \frac{v\sigma_R}{E} \qquad\qquad \text{(ii)}$$

Similarly a point at radius $(r + \delta r)$ moves radially outwards by an amount $(u + \delta u)$ and the radial strain of an element such as ABCD in Figure 5.4 is

$$\frac{\text{increase in radial width}}{\text{original radial width}} = \frac{[(r + \delta r + u + \delta u) - (r + u)] - \delta r}{\delta r}$$

$$= \frac{\delta u}{\delta r}$$

and, in terms of the corresponding stresses,

$$\frac{du}{dr} = \frac{\sigma_R}{E} - \frac{v\sigma_H}{E} \qquad\qquad \text{(iii)}$$

Obtaining du/dr from (ii) and substituting in (iii) we obtain:

$$\left(\frac{\sigma_H}{E} - \frac{v\sigma_R}{E}\right) + \frac{r}{E}\left(\frac{d\sigma_H}{dr} - \frac{vd\sigma_R}{dr}\right) = \frac{\sigma_R}{E} - \frac{v\sigma_H}{E}$$

$$(\sigma_H - \sigma_R)(1 + v) = r\left(v\frac{d\sigma_R}{dr} - \frac{d\sigma_H}{dr}\right)$$

Substituting for $\sigma_H - \sigma_R$ from (i),

$$\left(\rho\omega^2 r^2 + r\frac{d\sigma_R}{dr}\right)(1 + v) = r\left(v\frac{d\sigma_R}{dr} - \frac{d\sigma_H}{dr}\right)$$

from which

$$\frac{d\sigma_H}{dr} + \frac{d\sigma_R}{dr} = -\rho\omega^2 r(1+v)$$

Integrating with respect to r,

$$\sigma_H + \sigma_R = -\tfrac{1}{2}\rho\omega^2 r^2(1+v) + 2A \qquad\qquad\text{(iv)}$$

where the constant of integration is denoted by $2A$ for convenience later. Subtracting (i) from (iv) to eliminate σ_H,

$$2\sigma_R = -\tfrac{1}{2}\rho\omega^2 r^2(1+v) + 2A - \rho\omega^2 r^2 - r\frac{d\sigma_R}{dr}$$

$$2\sigma_R + r\frac{d\sigma_R}{dr} = -\tfrac{1}{2}\rho\omega^2 r^2(3+v) + 2A$$

The left-hand side can be written as $\dfrac{1}{r}\dfrac{d}{dr}(r^2\sigma_R)$ and hence

$$\frac{d}{dr}(r^2\sigma_R) = -\tfrac{1}{2}\rho\omega^2 r^3(3+v) + 2Ar$$

Integrating with respect to r

$$r^2\sigma_R = -\tfrac{1}{8}\rho\omega^2 r^4(3+v) + Ar^2 - B \qquad\qquad\text{(v)}$$

where the second integration constant is denoted by $-B$.
 On rearranging,

$$\sigma_R = A - \frac{B}{r^2} - \tfrac{1}{8}(3+v)\rho\omega^2 r^2$$

which is the required result. If it is substituted in (iv) the circumferential or hoop stress is found to be

$$\sigma_H = A + \frac{B}{r^2} - \tfrac{1}{8}(1+3v)\rho\omega^2 r^2$$

The radial stress σ_R is zero at a free surface and hence $\sigma_R = 0$ when $r = R_1$ and also when $r = R_2$. Thus

$$0 = A - \frac{B}{R_1^2} - \tfrac{1}{8}(3+v)\rho\omega^2 R_1^2$$

and

$$0 = A - \frac{B}{R_2^2} - \tfrac{1}{8}(3+v)\rho\omega^2 R_2^2$$

Solving these simultaneous equations for A and B we obtain

$$A = \tfrac{1}{8}\rho\omega^2(3+v)(R_1^2 + R_2^2)$$
$$B = \tfrac{1}{8}\rho\omega^2(3+v)(R_1^2 R_2^2)$$

Substituting the expression for σ_R the radial stress is given by

$$\sigma_R = \tfrac{1}{8}\rho\omega^2(3+v)\left(R_1^2 + R_2^2 - \frac{R_1^2 R_2^2}{r^2} - r^2\right)$$

Differentiating the expression in the last bracket with respect to r, and equating to zero to find the maximum,

$$2R_1^2 R_2^2/r^3 = 2r \quad \text{or} \quad r^2 = R_1 R_2$$

and the corresponding value of the expression in the bracket is

$$R_1^2 + R_2^2 - \frac{R_1^2 R_2^2}{R_1 R_2} - R_1 R_2 = (R_2^2 - 2R_1 R_2 + R_1^2)$$

$$= (R_2 - R_1)^2$$

Thus

$$\sigma_{R(\text{max})} = \tfrac{1}{8}\rho\omega^2(3+v)(R_2 - R_1)^2$$

With the numerical data given in the question,

$$\omega = 2\pi N/60 = 2\pi \times 3\,600/60 = 120\pi \text{ rad/s}$$

and

$$\begin{aligned}
\sigma_{R(\text{max})} &= \tfrac{1}{8} \times (7.5 \times 10^3 \text{ kg/m}^3) \times (120\pi \text{ rad/s})^2 \\
&\quad \times (3 + 0.33) \times [(30 - 7.5) \times 10^{-2} \text{ m}]^2 \\
&= 2.25 \times 10^7 \text{ N/m}^2 \quad \text{or} \quad 22.5 \text{ MN/m}^2 \qquad (Ans)
\end{aligned}$$

5.9 Radial and hoop stresses in thin rotating disc with small central hole

The general expressions for radial and hoop stresses at radius r in a thin circular rotating disc of uniform thickness are respectively

$$A + \frac{B}{r^2} - \tfrac{1}{8}(3+v)\rho\omega^2 r^2 \quad \text{and} \quad A - \frac{B}{r^2} - \tfrac{1}{8}(1+3v)\rho\omega^2 r^2$$

where A and B are constants, v = Poisson's ratio, ρ = density of the material and ω the angular velocity.

A thin disc of uniform thickness is 1 m external diameter and has a very small central hole. It rotates at 3 000 rev/min.

Calculate the maximum hoop stress deriving any formula used from those given above.

Take $\rho = 7.83 \text{ Mg/m}^3$ and $v = \tfrac{1}{3}$. (UL)

(Note The signs of B in the equations in the question are opposite to those of the previous problem. Since B is a constant of integration only the intermediate working is affected; the final results are unaltered.)

Solution As in the previous solution, $\sigma_R = 0$ when $r = R_1$ and also when $r = R_2$. From these boundary conditions,

$$A = \tfrac{1}{8}\rho\omega^2(3+v)(R_1{}^2 + R_2{}^2)$$

and

$$B = -\tfrac{1}{8}\rho\omega^2(3+v)(R_1{}^2 R_2{}^2)$$

With these values of A and B the hoop stress is

$$\sigma_H = \tfrac{1}{8}\rho\omega^2\left[(3+v)\left(R_1{}^2 + R_2{}^2 + \frac{R_1{}^2 R_2{}^2}{r^2}\right) - (1+3v)r^2\right]$$

From the form of this expression σ_H is greatest when r has its least value, i.e. R_1. Thus

$$\sigma_{H(\text{max})} = \tfrac{1}{4}\rho\omega^2[(1-v)R_1{}^2 + (3+v)R_2{}^2]$$

If the central hole is very small $(R_1 \approx 0)$ the maximum stress tends to the value

$$\sigma_{H(\text{max})} = \tfrac{1}{4}\rho\omega^2 R_2{}^2(3+v)$$

With the numerical data given in the question,

$$\omega = 2\pi\, N/60 = 2\pi \times 3\,000/60 = 100\pi \text{ rad/s}$$

and

$$\sigma_{H(\text{max})} = \tfrac{1}{4} \times (7.83 \times 10^3 \text{ kg/m}^3) \times (100\pi \text{ rad/s})^2 \times (1 \text{ m})^2 \times (3 + \tfrac{1}{3})$$
$$= 6.44 \times 10^8 \text{ N/m}^2 = 644 \text{ MN/m}^2 \qquad (Ans)$$

Problems

1 Solve the following thick cylinder problems using the assumptions of Example 5.1.

(a) If there is an internal pressure of 60 MN/m² and the hoop stress at the inner surface is 156 MN/m² (tensile), calculate the hoop stress at the outer surface and the longitudinal stress. The diameters are 120 mm and 80 mm.

(b) The pressure at the inside and outside edges of a tube are atmospheric (zero) and 30 MN/m² respectively. If the hoop stress at the inside edge is known to be 90 MN/m² (compressive) what is its value at the outside edge?

(c) At the outside surface of a thick cylinder the circumferential stress is known to be 22 MN/m² (tensile) when the internal pressure is 6 MN/m². Calculate the circumferential stress at the inside surface and at the point where the radial stress is 3.5 MN/m². If the diameters are 200 mm and 250 mm what is the longitudinal stress?

Answer (a) Required hoop stress, 96 MN/m²; longitudinal stress, 48 MN/m²; (b) 60 MN/m²; (c) Hoop stresses, 28 and 25.5 MN/m²; longitudinal stress, 11 MN/m².

2 Use the formulae derived in Examples 5.2 and 5.3 to solve the following problems. In (a), (b) and (e) draw diagrams showing the variation of the circumferential and radial stresses through the wall of the cylinder.

(a) The diameters of a thick-walled tube are 125 mm and 175 mm. Calculate the maximum and minimum hoop tensions due to an internal pressure of 80 bar.

(b) Calculate the maximum tensile and compressive hoop stresses in a cylinder which is simultaneously subjected to an internal pressure of 280 bar and an external pressure of 140 bar. The internal and external diameters are 125 mm and 225 mm respectively.

(c) Find the minimum external/internal diameter ratio for an internal pressure of 150 bar with a maximum hoop stress of 50 MN/m^2.

(d) Find the necessary thickness of a hydraulic main 150 mm internal diameter to withstand a pressure of 70 bar with a maximum hoop stress of 14 MN/m^2.

(e) Calculate the safe internal pressure in a cast-iron hydraulic cylinder, 450 mm internal diameter and 250 mm thick, if the safe tensile stress is 15 MN/m^2.

Answer (a) 24.67 and 16.67 MN/m^2; (b) 12.5 (tensile) and 1.5 (compressive) MN/m^2; (c) 1.36; (d) 54.9 mm; (e) 95.0 bar.

3 A thick-walled tube with closed ends is subjected to an internal pressure. If the external diameter is $k \times$ internal diameter prove that:

(a) The hoop stress at the inner surface is $(1 + k^2)/2$ times that at the outer surface.

(b) The hoop stress at the outer surface is twice the longitudinal stress.

(c) The ratio of the maximum hoop stress to the value given by thin cylinder theory is $(k^2 + 1)/(k + 1)$, taking d as the internal diameter in the thin cylinder case.

Use the result (c) to show that, when $k = 1.1$, thin cylinder theory underestimates the maximum hoop stress by about 5 per cent.

4 Show that the maximum (compressive) hoop stress in a cylinder subjected to an external pressure P is

$$-2Pk^2/(k^2 - 1)$$

where k is the external/internal diameter ratio.

The inner tube of a compound cylinder has internal and external diameters of 160 mm and 240 mm respectively. Calculate the external diameter of a second tube to be shrunk on to the first so that, due to the shrinkage, the maximum compressive

hoop stress in the inner tube is (numerically) equal to the maximum tensile hoop stress in the outer tube.
Answer 319 mm.

5 A hollow steel cylinder, 150 mm internal and 200 mm external diameter, is open at the ends and is subjected to an external pressure of 140 bar. Calculate the maximum compressive circumferential stress due to this pressure and the decrease in the internal and external diameters of the cylinder.

Take $E = 200 \, \text{GN/m}^2$ and $v = 0.28$.

Answer 64 MN/m²; 0.048 and 0.046 mm.

6 A compound tube has internal and external diameters of 150 mm and 250 mm respectively. It is formed by shrinking one tube on to another of the same material, the common diameter being 200 mm. If the shrinkage allowance is 0.05 mm calculate the radial pressure between the tubes at the common radius ($E = 200 \, \text{GN/m}^2$).

Find also the maximum and minimum hoop stresses in both tubes due to the shrinkage. If the compound tube is subjected to an internal pressure of 700 bar, what are the resultant stresses at the inner and outer surfaces of each tube?

Draw a diagram to show the distribution of the hoop stresses before and after the internal pressure is applied.

Answer Radial pressure at common radius, 6.15 MN/m². Hoop stresses in MN/m² (positive are tensile):

r (mm)	75	100 (inner)	100 (outer)	125
Due to shrinkage	−28.1	−22	28	21.9
Due to pressure	148.8	100.9	100.9	78.8
Resultant	120.7	78.9	128.9	100.7

7 Calculate for the compound tube of Problem 6 the minimum temperature difference between the two tubes to allow of the outer one passing over the inner. Coefficient of expansion, $\alpha = 11 \times 10^{-6}/°C$.

Find also the external diameter of a single tube to withstand the same internal pressure for the same maximum hoop stress (128.9 MN/m²), the internal diameter being 150 mm.
Answer 22.7°C; 276 mm.

8 A steel cylinder, 100 mm external diameter and 75 mm internal diameter, has another steel cylinder, external diameter 125 mm, shrunk on to it. If the maximum tensile hoop stress induced in the outer cylinder is 60 MN/m², find the radial compressive stress between the cylinders and the maximum compressive hoop stress in the inner cylinder.

By how much must the external diameter of the inner tube have exceeded the internal diameter of the outer tube to have produced these stresses? Take $E = 200\,\text{GN/m}^2$.

Answer $13.2\,\text{MN/m}^2$, $60.2\,\text{MN/m}^2$; $0.053\,5\,\text{mm}$.

9 Calculate the change in diameter of a $100\,\text{mm}$ solid shaft subjected to an external pressure of $80\,\text{MN/m}^2$. Take $E = 200\,\text{GN/m}^2$ and $v = 0.28$.

Answer $0.028\,8\,\text{mm}$ decrease.

10 A steel ring $150\,\text{mm}$ external diameter and $100\,\text{mm}$ internal diameter is shrunk on another steel ring having an internal diameter of $50\,\text{mm}$. The thickness of both rings is $20\,\text{mm}$. If the axial force required to push the rings apart is $56.54\,\text{kN}$ and the coefficient of friction for the mating surfaces is 0.2, find the pressure at the mating surfaces and the shrinkage allowance. Also sketch curves showing how the radial stress and the tangential hoop stress vary from the inside to the outside of the compound ring. Take $E = 200 \times 10^9\,\text{N/m}^2$.   (*IMechE*)

Answer $45\,\text{MN/m}^2$; $0.096\,\text{mm}$. Hoop stresses ($\sigma_H\,\text{MN/m}^2$) are:

r (mm)	25	50 (inner)	50 (outer)	75
σ_H	-120	-75	117	72

11 A steel plug, $75\,\text{mm}$ diameter, is forced into a steel ring, $125\,\text{mm}$ external diameter and $50\,\text{mm}$ wide. From a reading taken by fixing, in a circumferential direction, an electrical resistance strain gauge on the external surface of the ring the strain is found to be 1.5×10^{-4}. Assuming a coefficient of friction 0.2 for the mating surfaces, find the force required to push the plug out of the ring. Also estimate the greatest hoop stress in the ring. Take $E = 200\,\text{GN/m}^2$.   (*IMechE*)

Answer $62.8\,\text{kN}$; $56.7\,\text{MN/m}^2$.

12 A steel cylinder $200\,\text{mm}$ external diameter and $150\,\text{mm}$ internal diameter has another cylinder $250\,\text{mm}$ external diameter shrunk on to it.

If the maximum tensile stress induced in the outer cylinder is $75\,\text{MN/m}^2$, find the radial compressive stress between the cylinders.

Determine the circumferential stresses at the inner and outer diameters of both cylinders and show, by means of a diagram, how these stresses vary with the radius. Calculate the necessary shrinkage allowance at the common surface. Take $E = 200\,\text{GN/m}^2$.   (*UL*)

Answer   Radial pressure, $16.5\,\text{MN/m}^2$. Hoop stresses ($\sigma_H\,\text{MN/m}^2$) are:

r (mm)	75	100 (inner)	100 (outer)	125
σ_H	-75.25	-58.8	75	58.5

Shrinkage allowance $0.133\,\text{mm}$.

13 A solid circular shaft is subjected to a uniform radial compressive stress. Show, without using thick cylinder formulae, that the radial and circumferential stresses at all radii are equal to the external radial stress.

A steel shaft, originally 100 mm diameter, is subjected to a uniform radial compressive stress of 20 MN/m². Assuming the radial stress remains constant, find the uniform longitudinal stress required to reduce the initial diameter by 0.012 5 mm and calculate the alteration in volume for a 150 mm length of shaft.

Take $E = 200\,GN/m^2$ and Poisson's ratio $= 0.304$. (*UL*)

Answer 36.4 MN/m² (tensile); 8.24 mm³ (decrease).

14 Find the ratio of thickness to internal diameter for a tube subjected to internal pressure when the ratio of the internal pressure to the greatest circumferential stress is 0.5.

Find the alteration in thickness of metal in such a tube, 200 mm internal diameter, when the internal pressure is 750 bar. Take Poisson's ratio $= 0.304$, $E = 200\,GN/m^2$. (*UL*)

Answer 0.366; 0.012 5 mm (if longitudinal stress $= 0$).

15 A steel cylindrical plug of 125 mm diameter is forced into a steel sleeve of 200 mm external diameter and 100 mm long. If the greatest circumferential stress in the sleeve is 90 MN/m², find the torque required to turn the plug in the sleeve assuming the coefficient of friction between the plug and the sleeve is 0.2.

(*UL*)

Answer 19.36 kN m.

16 The inner tube of a compound cylinder is 15 cm external and 10 cm internal diameter. It is subjected to an external pressure of 40 MN/m² when the internal pressure is 120 MN/m². Working from first principles, determine the circumferential stress at the external and internal surfaces and find the radial and circumferential stresses at the mean radius. Draw a diagram to show how the radial and circumferential stresses vary with the radius.

(*UL*)

Answer Hoop stresses: at inner surface, 168 MN/m² (tensile); at outer surface 88 MN/m² (tensile). At mean radius: radial stress 68 MN/m² (compressive); hoop stress 116 MN/m² (tensile).

17 A steel shaft 40 mm diameter is to be encased in a bronze sleeve 60 mm outside diameter which is to be forced into position and, before forcing on, the inside diameter of the sleeve is smaller than the diameter of the shaft, the difference in these diameters being 0.05 mm.

Find, due to the forcing on,

(a) the radial pressure between the shaft and the sleeve,

(b) the maximum hoop stress in the sleeve,
(c) the change in the outside diameter of the sleeve.

For steel, $E = 200 \, \text{GN/m}^2$, Poisson's ratio $= 0.29$.
For bronze, $E = 120 \, \text{GN/m}^2$, Poisson's ratio $= 0.34$.   (UL)
Answer (a) $44.6 \, \text{MN/m}^2$; (b) $115.8 \, \text{MN/m}^2$ (tensile); (c) $0.021 \, 4 \, \text{mm}$ (increase).

18 A bronze bush having an outside diameter of 140 mm and an inside diameter of 80 mm is pressed into a recess in a body which is assumed to be perfectly rigid. If the diameter of the recess is 139.96 mm, find the radial pressure produced on the outer surface of the bush and the maximum hoop stress in the bush.
Determine also the change in the inside diameter of the bush.
For bronze take $E = 110 \, \text{GN/m}^2$ and Poisson's ratio $= 0.35$.
(UL)
Answer 19.4 and $57.6 \, \text{MN/m}^2$, $0.041 \, 9 \, \text{mm}$ (decrease).

19 The four elastic constants are E, G, K and v where E is Young's modulus of elasticity, G is the shear modulus of elasticity, K is the bulk modulus and v is Poisson's ratio.

(a) State the relationship between (i) E, G and v and (ii) E, K and v.
(b) A hollow cylinder having internal and external diameters of 90 mm and 150 mm was subjected to an internal pressure of $20 \, \text{MN/m}^2$ when the longitudinal strain was found to be 46×10^{-6}; this strain includes the effect of longitudinal stress. The same longitudinal strain was obtained by subjecting the cylinder to an axial pull of 59.3 kN with no internal pressure. Determine the values of the four elastic constants for the material from which the cylinder was made.   (UL)
Answer (a) (i) $E = 2G(1 + v)$. (ii) $E = 3K(1 - 2v)$. (b) $E = 114 \, \text{GN/m}^2$, $G = 45 \, \text{GN/m}^2$, $K = 81.5 \, \text{GN/m}^2$, $v = 0.267$.

20 State the general formulae for the hoop tension σ_H and the radial tension σ_R in a cylindrical pipe with a thick wall subjected to internal or external radial forces. From these formulae derive a formula for the hoop stress at a distance x from the axis of a hollow cylinder internal radius R_1 and external radius R_2 subjected to an internal pressure p_0.
A pipe of 18 mm external diameter and 6 mm bore carries oil at a pressure of $36 \, \text{MN/m}^2$. If the pipe is free of longitudinal stress, what is the maximum intensity of stress?   (UL)
Answer $\sigma_H = A + \dfrac{B}{x^2}$; $\sigma_R = A - \dfrac{B}{x^2}$ where A and B are constants and x is the distance from the axis;

$$\frac{p_0 R_1^2}{R_2^2 - R_1^2} \left(\frac{R_2^2}{x^2} + 1 \right); \quad 45 \, \text{MN/m}^2$$

21 A thick-walled steel cylinder having an inside diameter of 150 mm is to be subjected to an internal pressure of 40 MN/m². Find to the nearest mm the outside diameter required if the hoop tension in the cylinder wall is not to exceed 120 MN/m².

Using the diameter found above, calculate the actual hoop stress at the inner and outer surfaces of the cylinder and plot a graph showing how the hoop tension varies across the cylinder wall; calculate an intermediate value for this purpose. (*UL*)
Answer 213 mm; 118.8 and 78.8 MN/m².

22 A bronze sleeve having an outside diameter 75 mm is forced on to a steel rod 55 mm diameter, the initial inside diameter of the sleeve being 0.06 mm smaller than the rod diameter. When in service the compound rod is subjected to an external pressure of 20 MN/m² and at the same time to a rise in temperature of 100°C. Determine:

(a) the radial pressure between the sleeve and the rod; and
(b) the greatest circumferential stress in the sleeve under service conditions.

For steel, $E = 200$ GN/m², $v = 0.30$, $\alpha = 11 \times 10^{-6}$ per °C.
For bronze, $E = 100$ GN/m², $v = 0.33$, $\alpha = 19 \times 10^{-6}$ per °C.
(*UL*)

Answer Stresses in MN/m² (tensile positive).

	(a)	(b)
due to force fit	− 27.2	90.6
due to external pressure	− 21.6	−14.7
due to temperature rise	20.0	−66.4
resultant	− 28.8	9.5

23 A steel cylinder having inside and outside diameters of 200 mm and 275 mm respectively is shrunk on to another steel cylinder having inside and outside diameters of 150 mm and 200 mm respectively.

Find the necessary radial pressure between the cylinders due to shrinkage only so that when there is an internal pressure of 80 MN/m² the final value of the maximum hoop stress in the inner cylinder is 120 MN/m².

Determine also the magnitude of the maximum hoop stress in the outer cylinder when the internal pressure is acting. (*UL*)
Answer 6.07 MN/m²; 117.5 MN/m².

24 A steel cylinder is required to withstand an internal pressure of 500 bar, the outside diameter is to be 200 mm and the maximum allowable hoop tension is 125 MN/m².
Determine

(a) the greatest permissible value for the inside diameter;

(b) the change in the inside diameter due to the stresses in the cylinder wall;

(c) the maximum shearing stress in the cylinder wall.

$E = 200 \text{ GN/m}^2$, $v = 0.3$. (UL)

Hint Note that at radius r the maximum shearing stress is B/r^2 with the usual notation.

Answer (a) 131 mm; (b) 0.084 mm (allowing for longitudinal stress); (c) 87.5 MN/m².

25 A compound cylinder consists of two steel cylinders and has an outside diameter of 250 mm and an inside diameter of 150 mm; the common diameter between the cylinders is 200 mm. The compound cylinder is subjected to an internal pressure of 70 MN/m². The outer cylinder is shrunk on the inner cylinder and the initial difference between the common diameters is such that, when the internal pressure is applied, the maximum hoop stress in the inner cylinder is equal to the maximum hoop stress in the outer cylinder.

(a) Determine the radial pressure between the cylinders due to the shrink fit only.

(b) Calculate the initial difference between the common diameters.

(c) Make a diagram showing how the hoop stress varies across the cylinder wall when the internal pressure is applied.

For steel take $E = 203 \text{ GN/m}^2$. (UL)

Answer (a) 5.25 MN/m²; (b) 0.042 0 mm; (c) principal values

r (mm)	75	100 (inner)	100 (outer)	125
stress (MN/m²)	124.8	82.1	124.8	97.5

26 A hydraulic cylinder is designed to operate at a pressure of 600 bar with an internal diameter of 80 mm. Determine the minimum wall thickness required if the maximum hoop stress is not to exceed 250 MN/m².

By how much would the external diameter of the cylinder increase under this fully loaded condition? The cylinder is made of steel with $E = 200 \text{ GN/m}^2$ and $v = 0.26$. (UL)

Answer 11.1 mm; 0.084 5 mm.

27 A hub of 100 mm diameter and 100 mm long is shrunk on to a shaft of 50 mm diameter and is required to transmit a torque between the hub and shaft of 1 kN m. The coefficient of friction between the hub and shaft is 0.3. The radial stress on the outside of the hub is negligible.

Sketch the stress distribution across the thickness of the hub corresponding to the minimum required interface pressure and give values of the tangential stress at the inner and outer faces of the hub and of the radial stress at the inner face. (UL)

r (mm)	25	50
hoop stress (MN/m^2)	14.15	5.66
radial stress (MN/m^2)	8.49	0

28 Discuss the assumptions made in deriving the formulae for radial and hoop stress in pressurized 'thin' and 'thick' cylinders.

A simple thick cylinder is designed to withstand an internal pressure of 20 MN/m^2. It has an external diameter of 0.6 m and the maximum shear stress is 120 MN/m^2. Assuming that there is no external pressure determine (i) the thickness of the cylinder and (ii) the maximum hoop stress. If the actual thickness of a cylinder manufactured to this design is 20 per cent greater than the value determined from (i) above and there is an external compressive pressure of 6 MN/m^2, calculate the maximum internal pressure which the cylinder could withstand if the maximum shear stress is unaltered. *(UL)*

Answer (i) 26.1 mm; (ii) 220 MN/m^2; 30.8 MN/m^2.

29 Show that the maximum intensity of hoop tension in a thin rotating disc is

$$\tfrac{1}{8}\rho\omega^2 R^2(3 + v)$$

where ρ is the density of the material, R is the radius of the disc, ω its angular velocity and v is Poisson's ratio for the material.

(UL)

Hint In the usual equations for stress the constant B must be zero, otherwise the stress at $r = 0$ becomes infinite. (Note that the maximum stress in the solid disc is half that of the disc with a very small central hole, Example 5.9.)

30 A steel disc of 40 cm external diameter is of uniform thickness and has a 5 cm diameter hole at the centre. Find the speed of rotation about an axis perpendicular to the plane of the disc which will produce a maximum tensile stress of 80 MN/m^2.

Find also the value of the maximum radial stress and the radius at which it occurs.

The radial stress σ_x and the hoop stress σ_y at radius r are given by

$$\sigma_x = A - \frac{B}{r^2} - \tfrac{1}{8}\rho\omega^2 r^2(3 + v)$$

$$\sigma_y = A + \frac{B}{r^2} - \tfrac{1}{8}\rho\omega^2 r^2(1 + 3v)$$

where A and B are constants, ρ = density of material, ω = angular velocity, v = Poisson's ratio.

For steel $\rho = 7.83$ Mg/m^3 and $v = 0.3$. *(UL)*

Answer 5 305 rev/min; 30.52 MN/m^2 at 7.07 cm radius.

31 A thin circular steel disc of uniform thickness has an outside diameter of 50 cm with a central hole of 8 cm diameter and it rotates at a speed of 4 000 rev/min. Find (a) the maximum tensile stress in the disc, (b) the maximum shearing stress, and (c) the increase in the outside diameter caused by the rotation.

$E = 200 \text{ GN/m}^2$; $v = 0.28$; density of steel = 7.83 Mg/m³.

(*UL*)

Hint (a) the maximum tensile stress is the hoop stress at the inner radius; (b) the principal stresses at a given point are σ_H and σ_R and the maximum value of the corresponding shearing stress is $\frac{1}{2}(\sigma_H - \sigma_R)$, which has its greatest value at the inner radius; (c) the increase in diameter is proportional to the circumferential strain.

Answer (a) 70.8 MN/m²; (b) 35.4 MN/m²; (c) 0.043 mm.

32 A thick cylinder of inner diameter 120 mm and outer diameter 200 mm is subjected to an internal pressure of 100 MN/m². Determine the magnitude of the hoop stress at the inner radius.

A compound thick cylinder of the same overall dimensions is made by shrinking one tube over another of the same material having a common surface 160 mm so that the radial interface pressure at the common surface is initially 25 MN/m². Determine the hoop stress at the inner radius in this case when subject to the internal pressure of 100 MN/m². (*Brunel*)

Answer 212.5 MN/m²; 98.2 MN/m².

33 A thin steel disc of uniform thickness has an outside diameter of 550 mm and a central hole of 110 mm diameter. When it rotates at a speed of 5 000 rev/min a radial tensile stress of 80 MN/m² is applied at the outer circumference due to attached blading, while the bore is free of radial stress.

Find the value of the maximum tensile stress in the disc, and the increase in the inner and outer diameters compared with the disc at rest under no stress.

You may assume the formulae

$$\sigma_r = A + \frac{B}{r^2} - \left(\frac{3+v}{8}\right)\rho\omega^2 r^2$$

$$\sigma_\theta = A - \frac{B}{r^2} - \left(\frac{1+3v}{8}\right)\rho\omega^2 r^2$$

where the symbols have their usual meaning.

Take $E = 200 \text{ GN/m}^2$, $v = 0.3$ and the density of steel $\rho = 7\,830 \text{ kg/m}^3$. (*Brunel*)

Answer 301.7 MN/m²; 0.166 mm and 0.265 mm.

34 In order to shrink a steel bush 80 mm outer diameter and 50 mm long on to a solid steel shaft 50 mm diameter the bush is heated to 120°C at which temperature it is just possible to slide the bush on to the shaft which is at 20°C. Determine the radial pressure induced between the bush and shaft when cooled and hence estimate the torque which can be transmitted through the bush without slipping on the shaft, assuming a coefficient of friction of 0.3.

Take $E = 200\,GN/m^2$ and the linear coefficient of expansion of steel as 12×10^{-6} per °C. *(Brunel)*

Answer 187.2 MN/m²; 36.8 kN m.

6

Shearing stresses in beams

Variation of vertical shear stress

The vertical shear stress in a beam at a height h from the neutral axis is

$$\tau = \frac{F}{Ib_0} \int_h^Y yb \, dy = \frac{F}{Ib_0} A\bar{y}$$

where F = total shear force on the cross-section considered,
$\quad\quad I$ = second moment (or moment of inertia) of the area of the cross-section about the neutral axis,
$\quad\quad b_0$ = breadth of cross-section at the height h,
$\quad\quad b$ = breadth of cross-section at height y above the neutral axis,
$\quad\quad Y$ = distance from neutral axis to most strained fibre,
$\quad\quad A$ = area of cross-section of beam above height h,
$\quad\quad \bar{y}$ = distance of centroid of the area A from the neutral axis.

Maximum vertical shear stress

For a rectangular cross-section the ratio $\tau(\text{maximum})/\tau(\text{mean})$ is 3/2. The corresponding result for a circular cross-section is 4/3.

Approximation for I-beam

It is sufficiently accurate for some purposes to assume that the web carries the whole of the shearing force uniformly distributed over the area of the web.

Shear deflection

The shearing deflection over a short length δx of a beam or cantilever of rectangular section (width b and depth d) is

$$\frac{6F}{5Gbd} \delta x$$

where F is the local shearing force and G is the modulus of rigidity.
   The corresponding result for beams of other cross-sections should be derived when required from considerations of shear strain energy.

Shear centre

The loads carried by a beam or cantilever can produce torsion as well as bending. The shear centre is the point on the bending neutral axis through which the external loads must act if torsion is to be prevented.

Worked examples

6.1 Ratio of maximum to mean shear stresses in beam of symmetrical section

Working from first principles, derive a formula for calculating the intensity of shear stress at any point in the cross-section of a beam subjected to a transverse shearing force.

Find the ratio of the maximum shear stress to the mean shear stress for the beam section shown in Figure 6.1. (The centroid of a semicircle is $4R/3\pi$ from its centre.) *(IMechE)*

Solution Consider a length δx of the beam as shown in Figure 6.2(a).

Let M and $M + \delta M$ be the bending moments at the left- and right-hand ends of this portion respectively. From the bending equation, the longitudinal stress on a small area, distance y from the neutral axis (Figure 6.2(b)) at the left-hand end of this portion is

$$\sigma = \frac{My}{I}$$

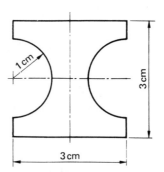

Figure 6.1

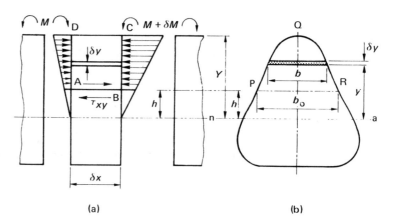

(a)   (b)

Figure 6.2

where I is the second moment of area of the cross-section. Hence the force on a slice width b, thickness δy and length δx pushing from left to right is

$$\sigma b \delta y = \frac{M y b \delta y}{I}$$

Similarly, from the conditions at the right-hand end there is a force pushing from right to left of

$$\frac{(M + \delta M) y b \delta y}{I}$$

Hence there is a resultant force from right to left on the slice of

$$\frac{(M + \delta M) y b \delta y}{I} - \frac{M y b \delta y}{I} = \frac{\delta M y b \delta y}{I} \tag{i}$$

Although the sum of all forces, such as (i), for the whole section is zero (there being no resultant longitudinal force) there is a resultant force if only part of the cross-section is considered. Consider the equilibrium of the block ABCD (Figure 6.2(a)) whose end section is PQR (Figure 6.2(b)).

Due to the bending moments there is a longitudinal force on this block given by the sum of all terms such as (i) for slices above AB. Since there can be no shear stress at a free surface such as CD, then the equilibrium of ABCD can only be maintained by a shear force along the face AB.

Assuming that the shear stress is uniform laterally and longitudinally, let its value be τ (in the direction shown). Then the total shear force acting on the block (from left to right) is

$$\tau \times AB \times PR = \tau \delta x b_0 \tag{ii}$$

where b_0 is the breadth of the section at AB.

For the equilibrium of ABCD the force given by (ii) must equal the sum of all forces such as (i) for the slices above AB. Hence

$$\tau \delta x b_0 = \sum \frac{\delta M y b \delta y}{I} = \frac{\delta M}{I} \sum y b \delta y \tag{iii}$$

δM and I being constants for any one section and the summation being considered over the whole area PQR.

From (iii),

$$\tau = \frac{\delta M}{\delta x} \frac{1}{I b_0} \sum y b \delta y$$

or, in the calculus notation,

$$\tau = \frac{dM}{dx} \frac{1}{I b_0} \int_h^Y y b \, dy$$

$$= \frac{F}{I b_0} \int_h^Y y b \, dy \tag{iv}$$

since $dM/dx = F$, the total shearing force. The limits of integration are h, the height of AB (and PR) above the neutral axis and Y, the distance from neutral axis to CD (or Q).

The integral $\int_h^Y yb\,dy$ is the first moment of the area PQR about the neutral axis. Hence, if A is the area PQR and $\bar{y}$ is the distance of its centroid from the neutral axis, (iv) can be rewritten as

$$\tau = \frac{F}{Ib_0} A\bar{y} \tag{v}$$

This longitudinal shearing stress is accompanied by a vertical (complementary) shearing stress of equal magnitude, and hence results (iv) and (v) are formulae for the vertical shearing stress at any height h from the neutral axis.

For the section given, the maximum shearing stress will occur at the neutral axis, since b_0 is a minimum there and $A\bar{y}$ has its maximum value there.

I for the section

$$= (I \text{ for a square } 3\,\text{cm} \times 3\,\text{cm}) - (I \text{ for a circle radius } 1\,\text{cm})$$

$$= \frac{3\,\text{cm} \times (3\,\text{cm})^3}{12} - \frac{\pi \times (2\,\text{cm})^4}{64}$$

$$= 5.965\,\text{cm}^4$$

Using the centroid position given in the question,

$A\bar{y} =$ (first moment of a rectangle $3\,\text{cm} \times 1\frac{1}{2}\,\text{cm}$ about longer edge)—(first moment of a semicircle, radius $1\,\text{cm}$, about bounding diameter)

$$= (3\,\text{cm} \times 1\tfrac{1}{2}\,\text{cm} \times \tfrac{3}{4}\,\text{cm}) - \left[\tfrac{1}{8}\pi \times (2\,\text{cm})^2 \times \frac{4 \times 1\,\text{cm}}{3\pi}\right]$$

$$= 2.708\,\text{cm}^3$$

Hence, the maximum shear stress is

$$\tau_{\text{max}} = \frac{F}{Ib_0} A\bar{y} = \frac{F \times 2.708\,\text{cm}^3}{5.965\,\text{cm}^4 \times 1\,\text{cm}}$$

$$= 0.454\,0F\,\text{N/cm}^2 = 4.540F\,\text{kN/m}^2$$

F being in newtons.

The mean shear stress is

$$\tau_{\text{mean}} = \frac{F}{\text{total area}} = F/[(3\,\text{cm} \times 3\,\text{cm}) - \tfrac{1}{4}\pi \times (2\,\text{cm})^2]$$

$$= 0.170\,7F\,\text{N/cm}^2 = 1.707F\,\text{kN/m}^2$$

The required ratio is

$$\frac{\tau_{\text{max}}}{\tau_{\text{mean}}} = \frac{4.540F}{1.707F} = 2.66 \tag{Ans}$$

Figure 6.3 shows the vertical cross-section of a beam which at this section is subjected to a vertical shearing force of 8 kN. Plot to a suitable scale a curve showing how the intensity of shear stress varies throughout the depth of the section. What is the ratio of the maximum shear stress to the mean shear stress?

(*IMechE*)

Solution　By the usual methods, the centroid, and hence the neutral axis, is 1.5 cm from the bottom edge and the corresponding second moment of area of the section is 8.667 cm^4.

Considering sections at each $\frac{1}{2}$ cm of the depth as shown in Figure 6.4, we have, from equation (v) of Example 6.1, the following results:

Section	A (cm^2)	$\bar{y}$ (cm)	b_0 (cm)	$\tau = \dfrac{F}{Ib_0} A\bar{y}$ (MN/m^2)
AA	0	—	—	0
BB	0.5	2.25	1	10.39
CC	1.0	2.00	1	18.46
DD	1.5	1.75	1	24.23
EE ($b_0 = 1$ cm)	2.0	1.50	1	27.69
FF ($b_0 = 3$ cm)	2.0	1.50	3	9.23
n.a. (neutral axis)	4.5	0.75	3	10.39
GG	3.0	1.0	3	9.23
HH	1.5	1.25	3	5.77
JJ	0	—	—	0

The last four cases have been calculated by considering the areas between the sections and the bottom edge.

Two values (at EE and FF) have been calculated for the 'junction' between the top and bottom portions since theoretically there is a sudden change in the shear stress intensity at this point. Since there

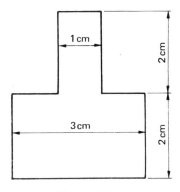

Figure 6.3

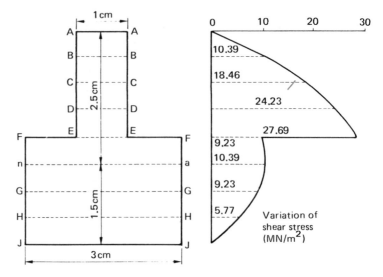

Figure 6.4

can be no shear stress along the top edge of the bottom portion (except where it joins the top portion) the theory fails to give the correct values at such sections. A similar case occurs with the underside of a flange or the inside faces of a hollow beam. In practice, there is a lateral variation of shear stress.

The values obtained are plotted in Figure 6.4 alongside the section view.

The mean shearing stress for the section is

$$\tau_{mean} = \frac{F}{\text{total area}} = \frac{8\,\text{kN}}{(3\,\text{cm} \times 2\,\text{cm}) + (1\,\text{cm} \times 2\,\text{cm})}$$

$$= 1\,\text{kN/cm}^2 = 10\,\text{MN/m}^2$$

The maximum shear stress clearly occurs at EE and hence the required ratio is

$$\frac{\tau_{max},}{\tau_{mean}} = \frac{27.69\,\text{MN/m}^2}{10\,\text{MN/m}^2} = 2.769 \qquad (Ans)$$

6.3 Variation of shear stress in beams of rectangular and circular cross-sections

Establish formulae for the shear stress at a distance h from the neutral axis at a section of a beam where the total shear force is F, if the cross-section of the beam is

(a) a rectangle, width b and depth d,
(b) a circle, diameter D.

Find, in each case, the ratio of the maximum to the mean shear stress and the distance from the neutral axis at which the local shear stress equals the mean shear stress.

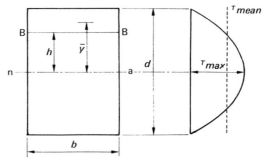

Figure 6.5

Solution

(a) Let BB (Figure 6.5) be the required distance h from the neutral axis. Then the area of the section above BB is

$$A = b(\tfrac{1}{2}d - h)$$

The distance of its centroid from the neutral axis is

$$\bar{y} = \tfrac{1}{2}(\tfrac{1}{2}d + h)$$

Using equation (v) of Example 6.1, the shear stress at XX is

$$\tau = \frac{F}{Ib_0} A\bar{y} = \frac{F}{(bd^3/12) \times b} \times b(\tfrac{1}{2}d - h) \times \tfrac{1}{2}(\tfrac{1}{2}d + h)$$

$$= \frac{6F}{bd^3}[(\tfrac{1}{2}d)^2 - h^2] \qquad\qquad\qquad (i)$$

At the top and bottom edges $h^2 = (\tfrac{1}{2}d)^2$ and $\tau = 0$.

The maximum value of τ occurs at the neutral axis when $h = 0$ and

$$\tau_{max} = \frac{6F}{bd^3} \times (\tfrac{1}{2}d)^2 = \frac{3F}{2bd}$$

The mean shear stress is

$$\tau_{mean} = \frac{F}{\text{total area}} = \frac{F}{bd}$$

Hence the required ratio is

$$\frac{\tau_{max}}{\tau_{mean}} = \left(\frac{3F}{2bd}\right)\bigg/\left(\frac{F}{bd}\right) = \frac{3}{2} \qquad\qquad (Ans)$$

The local shear stress (τ) equals the mean (τ_{mean}) where

$$\frac{6F}{bd^3}[(\tfrac{1}{2}d)^2 - h^2] = \frac{F}{bd}$$

$$\tfrac{1}{4}d^2 - h^2 = \tfrac{1}{6}d^2 \qquad h^2 = d^2/12$$

$$h = d/\sqrt{12} = 0.289\,d \qquad\qquad\qquad (Ans)$$

From (i) the variation of τ is parabolic and is shown in Figure 6.5.

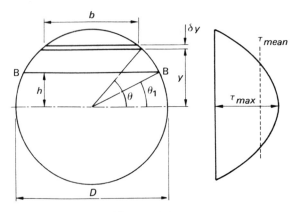

Figure 6.6

(b) Since there are no simple formulae for the area and centroid position of a segment of a circle, equation (iv) of Example 6.1 will be used. Let BB (Figure 6.6) be the required distance h above the neutral axis. The integration is most easily effected 'by expressing b and y in terms of the angle θ shown in Figure 6.6.

$$y = \tfrac{1}{2}D \sin \theta \quad \text{hence} \quad \frac{dy}{d\theta} = \tfrac{1}{2}D \cos \theta$$

Therefore, dy may be replaced by $\tfrac{1}{2}D \cos \theta \, d\theta$ if the limits of integration are also written in terms of angles. Let θ_1 be the value of θ when $y = h$. At the top of the section where $y = \tfrac{1}{2}D$, $\theta = \tfrac{1}{2}\pi$. Also,

$$b = 2 \times \tfrac{1}{2}D \cos \theta = D \cos \theta \quad \text{and} \quad b_0 = D \cos \theta_1.$$

Substituting in equation (iv) of Example 6.1, the shear stress at the required height is

$$\tau = \frac{F}{Ib_0} \int_h^Y yb \, dy$$

$$= \frac{F}{(\pi D^4/64)D \cos \theta_1} \int_{\theta_1}^{\pi/2} (\tfrac{1}{2}D \sin \theta) \times (D \cos \theta)\tfrac{1}{2}D \cos \theta \, d\theta$$

$$= \frac{16F}{\pi D^2 \cos \theta_1} \int_{\theta_1}^{\pi/2} \cos^2 \theta \sin \theta \, d\theta$$

$$= \frac{16F}{\pi D^2 \cos \theta_1} \int_{\theta_1}^{\pi/2} (-\cos^2 \theta) \, d(\cos \theta)$$

$$= \frac{16F}{\pi D^2 \cos \theta_1} \left[-\frac{\cos^3 \theta}{3} \right]_{\theta_1}^{\pi/2} = \frac{16F \cos^3 \theta_1}{3\pi D^2 \cos \theta_1}$$

$$= \frac{16F}{3\pi D^2} (1 - \sin^2 \theta_1) = \frac{16F}{3\pi D^2} \left[1 - \left(\frac{2h}{D}\right)^2 \right] \tag{ii}$$

(since $h = \tfrac{1}{2}D \sin \theta_1$).

The graph given by (ii) is a parabola as shown in Figure 6.6. The maximum value of τ occurs at the neutral axis where $h = 0$, and is

$$\tau_{max} = \frac{16F}{3\pi D^2}$$

The mean shear stress is

$$\tau_{mean} = \frac{F}{\text{total area}} = \frac{4F}{\pi D^2}$$

Hence the required ratio is

$$\frac{\tau_{max}}{\tau_{mean}} = \left(\frac{16F}{3\pi D^2}\right) \Big/ \left(\frac{4F}{\pi D^2}\right) = \frac{4}{3} \qquad (Ans)$$

The local shear stress (τ) equals the mean (τ_{mean}) where

$$\frac{16F}{3\pi D^2}\left[1 - \left(\frac{2h}{D}\right)^2\right] = \frac{4F}{\pi D^2}$$

$$1 - \left(\frac{2h}{D}\right)^2 = \frac{3}{4}$$

$$\left(\frac{2h}{D}\right) = \frac{1}{2}$$

$$h = \tfrac{1}{4}D \qquad (Ans)$$

6.4 Variation of shear stress in beam of I-section

A beam of I-section, 20 cm deep and 8 cm wide, has flanges 1 cm thick and web $\frac{1}{2}$ cm thick. It carries at one section a shearing force of 30 kN. Calculate the shear stress in the flange at $\frac{1}{2}$ cm and 1 cm from the outside edges.

Find an expression for the shear stress in the web at any distance h cm from the neutral axis, and draw a diagram showing the variation of shear stress across the section.

What percentage of the total shear force is carried by the web? What is the intensity of shear stress in the web, assuming that the web carries the whole shearing force uniformly distributed?

Solution The section is shown in Figure 6.7(a) and, by the usual methods, the second moment of area about the neutral axis is $I = 1688$ cm^4.

The shear stress at a point in the flange can be found from equation (v) of Example 6.1.

At $\frac{1}{2}$ cm from outside edge,

$$A = \tfrac{1}{2} \text{ cm} \times 8 \text{ cm} = 4 \text{ cm}^2$$

$$\bar{y} = 10 \text{ cm} - \tfrac{1}{4} \text{ cm} = 9\tfrac{3}{4} \text{ cm}$$

$$b_0 = 8 \text{ cm}$$

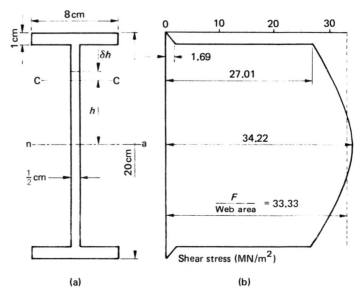

Figure 6.7

and hence

$$\tau = \frac{F}{Ib_0}A\bar{y} = \frac{30\text{ kN} \times 4\text{ cm}^2 \times 9\tfrac{3}{4}\text{ cm}}{1\,688\text{ cm}^4 \times 8\text{ cm}}$$

$$= 0.0866\text{ kN/cm}^2 = 0.866\text{ MN/m}^2 \qquad (Ans)$$

At 1 cm from outside edge

$$A = 1\text{ cm} \times 8\text{ cm} = 8\text{ cm}^2$$

$$\bar{y} = 10\text{ cm} - \tfrac{1}{2}\text{ cm} = 9\tfrac{1}{2}\text{ cm}$$

$$b_0 = 8\text{ cm}$$

Hence

$$\tau = \frac{F}{Ib_0}A\bar{y} = \frac{30\text{ kN} \times 8\text{ cm}^2 \times 9\tfrac{1}{2}\text{ cm}}{1\,688\text{ cm}^4 \times 8\text{ cm}}$$

$$= 0.169\text{ kN/cm}^2 = 1.69\text{ MN/m}^2$$

In the web at a distance h cm from the neutral axis the shear stress can be found by equations (iv) or (v) of Example 6.1. In either case the area above CC (Figure 6.7) must be considered in two parts, the flange and that part of the web above CC.

Using the integral method, with kN and cm units,

$$\tau = \frac{F}{Ib_0}\int_h^Y yb \cdot dy = \frac{30}{1\,688 \times \tfrac{1}{2}}\left\{\int_h^9 y \cdot \tfrac{1}{2} \cdot dy + \int_9^{10} y \cdot 8 \cdot dy\right\}$$

$$= \frac{30}{844}\left\{\left[\frac{y^2}{4}\right]_h^9 + \left[4y^2\right]_9^{10}\right\} = \frac{30}{844}[(20\tfrac{1}{4} - \tfrac{1}{4}h^2) + (400 - 324)]$$

$$= (34.22 - 0.088\,9h^2)\text{ MN/m}^2$$

Using the first moment of area method,

$$A\bar{y}(\text{total}) = A\bar{y}(\text{flange}) + A\bar{y}(\text{that part of the web above CC})$$
$$= [8 \times 1 \times 9\tfrac{1}{2}] + [\tfrac{1}{2}(9 - h) \times \tfrac{1}{2}(9 + h)]$$
$$= 76 + \tfrac{1}{4}(81 - h^2) = 96\tfrac{1}{4} - \tfrac{1}{4}h^2$$

Hence

$$\tau = \frac{F}{Ib_0}A\bar{y} = \frac{30}{1\,688 \times \tfrac{1}{2}}(96\tfrac{1}{4} - \tfrac{1}{4}h^2)\text{ kN/cm}^2$$
$$= (34.22 - 0.088\,9h^2)\text{ MN/m}^2$$

Using this equation for τ the following values are obtained,

h cm	0	± 3	± 6	± 9
τ MN/m^2	34.22	33.42	31.02	27.01

A diagram showing the variation of shear stress across the section is given in Figure 6.7(b).

The shear *force* carried by an element of the web, thickness δh (at CC) as shown in Figure 6.7(a), is $\tau \times \delta h \times b$. Using the expression for τ obtained above and working again in kN and cm units, the total shear force carried by the web is

$$\int_{-9}^{+9} \tau b\, dh = 2\int_0^9 (3.422 - 0.008\,89h^2)\tfrac{1}{2}dh$$
$$= \left[3.422h - 0.008\,89\left(\frac{h^3}{3}\right)\right]_0^9$$
$$= 28.64\text{ kN}$$

Hence, the percentage of the total shear force carried by the web is

$$\frac{28.64}{30} \times 100 \quad \text{i.e. } 95.5 \text{ per cent} \tag{Ans}$$

If the total shear force were uniformly distributed over the web the shear stress would be

$$\tau = \frac{30\text{ kN}}{18\text{ cm} \times \tfrac{1}{2}\text{ cm}} = 3.33\text{ kN/cm}^2 = 33.3\text{ MN/m}^2 \tag{Ans}$$

(This result is a good approximation to the shear stress at all points in the web and, in practice, it is often assumed that the web carries the total shear force uniformly distributed.)

6.5 Principal stress in I-section beam due to bending and shearing

A horizontal beam of I section 10 cm deep has flanges 6 cm by 0.7 cm and a web 0.36 cm thick. The transverse loading is such that at a certain section there is a bending moment of 3 kN m together with a vertical shearing force. If the maximum principal stress is limited to 80 MN/m^2 what is the value of the shearing force? (*IMechE*)

Solution By the usual methods, the second moment of area of the section is $I = 201.1 \text{ cm}^4$. The bending stress varies (linearly) from zero at the neutral axis to a maximum at the outside edge of the flanges. The shear stress variation is similar to that shown in Figure 6.7(b). The shear stress is small in the flange where the bending stress is greatest and is a maximum at the neutral axis where the bending stress is zero. The maximum principal stress occurs in the web at the point where it joins the flange, since the bending stress there is nearly equal to that at the outside edge and the shear stress is little short of the value at the neutral axis.

If F kN is the vertical shearing force on the section the shear stress in the web at this point (4.3 cm from the neutral axis) is

$$\tau = \frac{F}{I b_0} A\bar{y}$$

$$= \frac{F \text{ kN}}{201.1 \text{ cm}^4 \times 0.36 \text{ cm}} [6 \text{ cm} \times 0.7 \text{ cm} \times (5 - 0.36) \text{ cm}]$$

$$= 0.269\,8F \text{ kN/cm}^2 = 2.698F \text{ MN/m}^2$$

The bending stress at the same point is

$$\sigma = \frac{My}{I} = \frac{3 \text{ kN m} \times 4.3 \text{ cm}}{201.1 \text{ cm}^4} = 6.415 \text{ kN/cm}^2$$

$$= 64.15 \text{ MN/m}^2$$

With the usual relationship for principal stresses

$$(\sigma - \sigma_x)(\sigma - \sigma_y) = \tau_{xy}^2$$

In the present case,

the principal stress $\sigma = 120 \text{ MN/m}^2$

the bending stress $\sigma_x = 64.15 \text{ MN/m}^2$,

$\sigma_y = 0$ and $\tau_{xy} = (2.698F) \text{ MN/m}^2$.

Hence

$$(80 - 64.15)(80 - 0) = (2.698F)^2$$

from which

$$F^2 = 174.2 \qquad F = 13.2$$

The shearing force is therefore 13.2 kN. *(Ans)*

6.6 Deflections of I-section beam due to shearing and bending

The beam of Example 6.4 is simply supported over a span of 6 m. Find, for a central concentrated load of 50 kN the maximum deflection

(a) due to shearing, assuming the web carries the whole shearing force uniformly distributed
(b) due to bending.

Take $E = 200 \text{ GN/m}^2$ and $G = 80 \text{ GN/m}^2$.

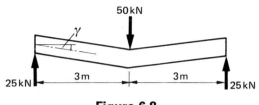

Figure 6.8

Solution (a) Figure 6.8 shows the type of distortion due to shear assuming uniform shear stress in the web,

Shear stress in web is

$$\tau = \frac{\text{shear force}}{\text{web area}} = \frac{25 \text{ kN}}{18 \text{ cm} \times \frac{1}{2} \text{ cm}} = 27.78 \text{ MN/m}^2$$

and the corresponding shear strain is

$$\gamma = \frac{\tau}{G} = \frac{27.78 \text{ MN/m}^2}{80 \text{ GN/m}^2} = 0.000\,347\,3$$

Hence, the mid-span deflection due to shear is

$$\gamma \times 3 \text{ m} = 0.000\,347\,3 \times 3$$
$$= 0.001\,04 \text{ m} \quad \text{or} \quad 1.04 \text{ mm} \qquad (Ans)$$

(b) With the usual formula, the deflection due to bending is

$$\delta = \frac{WL^3}{48EI} = \frac{50 \text{ kN} \times (6 \text{ m})^3}{48 \times 200 \text{ GN/m}^2 \times 1\,688 \text{ cm}^4}$$
$$= 0.066\,74 \text{ m} \quad \text{or} \quad 66.74 \text{ mm} \qquad (Ans)$$

(In most practical beams and cantilevers the shear deflection is small compared with that due to bending.)

6.7 Deflection of cantilever of rectangular section due to shearing and bending

Obtain expressions for the deflection at the free end of a cantilever, of rectangular cross-section, due to shearing strain when it carries

(a) a concentrated load at the free end
(b) a uniformly distributed load

Calculate the ratio (shear deflection)/(bending deflection) in each case if $E = 2.5G$ and the length of the cantilever is 8 times the depth.

Solution The required results can be derived by strain energy methods. Referring to Figure 6.2, consider a small slice length δx and width b_0 (represented by AB and PR) and of thickness δh. Its volume is $b_0 \delta x \delta h$ and if τ is the shear stress then the corresponding shear strain energy δU is given by

$$\delta U = \frac{\tau^2}{2G} \times \text{volume} = \frac{\tau^2}{2G} b_0 \delta x \delta h \qquad \text{(i)}$$

and, in general, τ varies with both x and h.

(a) If the cantilever carries a concentrated load W at the free end then the shearing force F is constant along its length and $F = W$. Thus, from equation (i), of Example 6.3 we have, for a rectangular cross-section,

$$\tau = \frac{6W}{bd^3}[(\tfrac{1}{2}d)^2 - h^2]$$

Substituting this result in (i) above, gives

$$\delta U = \frac{1}{2G}\left(\frac{6W}{bd^3}\right)^2 [(\tfrac{1}{2}d)^2 - h^2]^2 b_0 \delta x \delta h$$

Since there is no variation along the cantilever we may replace δx with L (total length) and integrate with respect to h. Thus since $b_0 = b$ (constant) we have

$$U = \frac{L}{2G}\int_{-d/2}^{+d/2}\left(\frac{6W}{bd^3}\right)^2 [(\tfrac{1}{2}d)^2 - h^2]^2 b\,dh$$

$$= \frac{18LW^2}{Gbd^6}\left[\frac{d^4 h}{16} - \frac{d^2 h^3}{6} + \frac{h^5}{5}\right]_{-d/2}^{+d/2}$$

$$= \frac{3W^2 L}{5Gbd}$$

If δ is the deflection at the free end due to shearing strain then the work done by the load W in producing this deflection is $\tfrac{1}{2}W\delta$. Equating this to the total internal shear strain energy gives

$$\tfrac{1}{2}W\delta = \frac{3W^2 L}{5Gbd} \quad \text{or} \quad \delta = \frac{6WL}{5Gbd} \qquad \text{(ii)} \quad \text{(Ans)}$$

The deflection due to bending is $WL^3/3EI$ and with $E = 2.5G$ and $L = 8d$ (as given in the question) the required ratio is

$$\frac{\delta(\text{shearing})}{\delta(\text{bending})} = \left(\frac{6WL}{5Gbd}\right) \bigg/ \left(\frac{WL^3}{3EI}\right) = \frac{18EI}{5GbdL^2}$$

$$= \frac{18E}{5GbdL^2}\left(\frac{bd^3}{12}\right) = \frac{3E}{10G}\left(\frac{d}{L}\right)^2$$

$$= \frac{3}{10} \times 2.5 \times \left(\frac{1}{8}\right)^2 = \frac{3}{256} \qquad \text{(Ans)}$$

(b) In this case there is a variation of shearing force along the cantilever. Consider a small portion, length δx, distance x from the fixed end as shown in Figure 6.9. This element can be considered as a

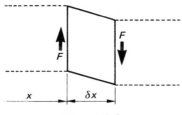

Figure 6.9

cantilever with an end load equal to the local shearing force F. The shearing deflection of this element can be obtained from expression (ii) above replacing W by F and L by δx. The deflection of the element is therefore $(6F/5Gbd)\delta x$ and, for the whole cantilever, the deflection

$$\delta = \int_0^L \frac{6F}{5Gbd} \, dx \qquad \text{(iii)}$$

For a uniformly distributed load w per unit length the shearing force at distance x from the fixed end is $F = w(L - x)$ and (iii) becomes

$$\delta = \frac{6}{5Gbd} \int_0^L w(L - x) \, dx$$

$$= \frac{6w}{5Gbd} [Lx - \tfrac{1}{2}x^2]_0^L$$

$$= \frac{3wL^2}{5Gbd} = \frac{3WL}{5Gbd} \qquad \text{(Ans)}$$

where $W = wL$, the total load.

On comparison with the bending deflection $(WL^3/8EI)$ we have

$$\frac{\delta(\text{shearing})}{\delta(\text{bending})} = \left(\frac{3WL}{5Gbd}\right) \bigg/ \left(\frac{WL^3}{8EI}\right)$$

$$= \frac{24EI}{5GbdL^2} = \frac{2E}{5G} \left(\frac{d}{L}\right)^2$$

If $E = 2.5G$ and $L = 8d$, this gives a ratio of 1/64.   (Ans)
(In both (a) and (b) the shearing deflection is small compared with that due to bending.)

6.8 Shear centre for beam of channel section

Figure 6.10 shows the section of a horizontal cantilever which carries a vertical end load W. Determine the position of W relative to the edge AC in order that the cantilever shall not be subjected to torsion. Work from first principles.   (UL)

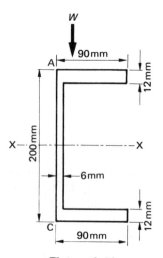

Figure 6.10

Solution In a beam or cantilever of open section such as a T, I or channel section it is found (see Example 6.4) that the vertical web carries nearly all the shearing force and the vertical shearing stresses in the flanges are small. However, horizontal shearing stresses occur in the flanges and torsion may result.

Suppose, in Figure 6.11(a), PQRS is a small element of the flange of a beam or cantilever, length δx and end area A. Let M and $M + \delta M$ be the bending moments at the sections SR and PQ respectively. Then if $\bar{y}$ is the distance of the centroid of the area A from XX, there is a net longitudinal force on the element due to the bending moments of

$$\frac{\delta M \bar{y}}{I} \times A \qquad (i)$$

As shown in Figure 6.11(b) this can only be resisted by a shearing stress τ acting along the face PS (since there can be no shearing stress

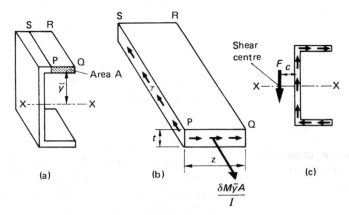

Figure 6.11

on a free surface such as QR). If t is the flange thickness at P this force is

$$\tau \times \text{area} = \tau \times t\delta x \tag{ii}$$

Equating (i) and (ii) we obtain

$$\tau = \frac{A\bar{y}}{It} \times \frac{\delta M}{\delta x}$$

In the limiting case $dM/dx = F$, the shearing force, and

$$\tau = \frac{FA\bar{y}}{It} \tag{iii}$$

Since complementary shear stresses are equal there is a horizontal shear stress τ in the flange at P and the general direction of the shear stress on the section is everywhere parallel to the boundary as shown in Figure 6.11(c).

In a channel section the resulting forces in the flanges form a couple and torsion will occur unless an equal and opposite couple is produced by the external force and the shear force in the web. The line along which the external force must act to prevent torsion is independent of the magnitude of the force and its point of intersection with XX is known as the *shear centre*.

For the section shown in Figure 6.10 and working in mm units, $I_{xx} = 21.8 \times 10^6 \text{ mm}^4$. At a point in the flange such as P, distance z from the edge Q, the area $A = tz$, $\bar{y} = 94$ mm and, from (iii),

$$\tau = \frac{F \times tz \times 94}{21.8 \times 10^6 t} = 4.31 \times 10^{-6} Fz$$

By integration the total force in one flange is

$$\int_0^{84} \tau t \, dz = \int_0^{84} 4.31 \times 10^{-6} Fz \times 12 \, dz$$

$$= 51.72 \times 10^{-6} F \left[\frac{z^2}{2} \right]_0^{84}$$

$$= 0.182\,5F$$

There is an equal and opposite force in the lower flange and these two forces are 188 mm apart. If the shear centre is distance c from the centre-line of the web, see Figure 6.11(c), then, equating the moments of the couples,

$$Fc = 0.182\,5F \times 188 \qquad c = 34.3 \text{ mm}$$

Hence the load W must act along a line 31.3 mm (outwards) from the edge AC. (*Ans*)

Problems

1 A cantilever of rectangular cross-section 90 mm deep and 30 mm wide carries a concentrated load at the free end. Calculate the value of this load if

(a) the mean (vertical) shear stress is 60 MN/m², and
(b) the maximum (vertical) shear stress is to be 60 MN/m².

Determine the diameter of a cantilever of circular cross-section to carry the same load as the rectangular cantilever above with

(c) the same mean shear stress,
(d) the same maximum shear stress.

Answer (a) 162 kN; (b) 108 kN; (c) 58.6 mm; (d) 55.3 mm.

2 Show that, according to the usual theory, the ratio of the maximum shear stress to the mean shear stress on a section of a hollow circular beam is $4(D^2 + Dd + d^2)/3(D^2 + d^2)$, where $D =$ external diameter and $d =$ internal diameter.

Hence show that for all values of D/d, the ratio lies between $\frac{4}{3}$ and 2. For what value of D/d is the ratio $\frac{5}{3}$?
Answer 3.73.

3 A bar of square section is used as a beam so that the plane of bending is parallel to a diagonal. The side of the square is 40 mm and the shearing force at one section is 16 kN.
Calculate

(a) the mean shear stress,
(b) the shear stress at the neutral axis,
(c) the magnitude and position of the maximum shear stress.

Draw a diagram showing the variation of shear stress across the section.
Answer (a) 10 MN/m²; (b) 10 MN/m²; (c) 11.25 MN/m² at 7.07 mm from neutral axis. The variation is shown in Figure 6.12.

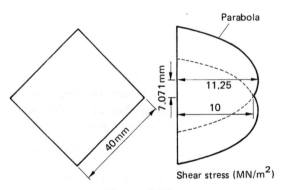

Figure 6.12

4 A beam has the section shown in Figure 6.13 and is used with the cross-piece horizontal and uppermost. It carries at one section a total (vertical) shear force of 15 kN. Draw to scale a diagram showing the distribution of shear stress across this section. Prove the formula giving the shear stress at any section.
Answer At h mm from the top edge the shear stress τ MN/m^2 is given by

h	0	10	20	20	25	30	40	60	80
			(flange)	(web)					
τ	0	3.69	5.53	27.67	27.89	27.67	25.81	16.60	0

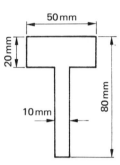

Figure 6.13

5 A beam of I section, 6 m long, has an overall depth of 254 mm, the flanges are 152 mm wide and 25.4 mm thick and the web thickness is 12.7 mm $(I = 110.4 \times 10^6 \text{ mm}^4)$. If the beam carries a uniformly distributed load of 400 kN total and is simply supported at its ends, calculate the maximum (vertical) shear stresses in the web and flanges at a section 1.2 m from one end. Find also the percentage of the total shear force at any section carried by the web.
Answer Maximum shear stress in web, 43.38 MN/m^2; in flange, 3.16 MN/m^2. Web carries 89.2 per cent.

6 The cantilever bracket, shown in Figure 6.14 is pulled with force P, equal to 14.14 kN, inclined at 45° to the horizontal. Find, for the point A,

  (a) the intensity of shear stress,
  (b) the intensity of normal stress on the cross-section, and
  (c) the principal stresses. *(IMechE)*

Hint Resolve P into its vertical and horizontal components. The vertical component gives a bending moment and shearing force. The horizontal component is an 'eccentric load'.
Answer (a) 16.4 MN/m^2; (b) 221.9 MN/m^2; (c) 223.2 MN/m^2 (compressive) and 1.2 MN/m^2 (tensile).

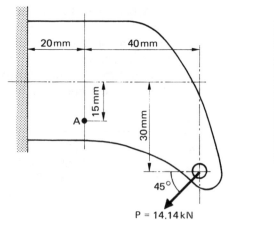

Figure 6.14

7 If the section (b m $\times$ d m) of a rectangular beam is subjected to a shearing force of V N, show that the shear stress at s metres above the neutral axis is

$$\frac{6V}{bd^3}\left\{\left(\frac{d}{2}\right)^2 - s^2\right\} \text{N/m}^2$$

Determine the principal stresses (at 0.013 m down from the upper face) at the built-in end of a cantilever of length 0.3 m and cross-section $b = 0.02$ m, $d = 0.04$ m. The cantilever carries 500 N at its free end and 250 N at 0.15 m from the free end.   (*UL*)
Answer   12.42 MN/m² (tensile); 0.12 MN/m² (compressive).

8 A bar of hexagonal cross-section and of side length 10 mm is used as a cantilever, one diagonal being horizontal. A load hung from the bar subjects it to a shearing force of 2 kN. Plot the shear stress distribution diagram.   (*UL*)
Answer   The shear stress τ MN/m² at h mm from the neutral axis is given by

h	0	±1.68	±2	±4	±6	±8	±8.66
τ	9.24	9.69 (max)	9.67	8.76	6.31	1.96	0

9 A steel cantilever of I-section 20 cm deep and 8 cm wide has flanges 1 cm thick and web $\frac{1}{2}$ cm thick. It is 2 m long and carries at the free end a concentrated load of 10 kN.
Calculate, neglecting the self-weight of the cantilever,

(a) the maximum (vertical) shear stress in the web,
(b) the deflection at the free end due to bending,

(c) the deflection at the free end due to shear, assuming the web carries the whole of the shear force uniformly distributed.

$E = 200 \times 10^9 \text{ N/m}^2$; $G = 80 \times 10^9 \text{ N/m}^2$.

Answer (a) 11.4 MN/m²; (b) 8.0 mm; (c) 0.278 mm.

10 A compound girder has a top flange 6 cm wide by 1 cm thick, the bottom flange being similarly 12 cm by 1 cm, and the web is $\frac{1}{2}$ cm thick, the girder being 50 cm deep overall. At a certain section the shearing force is 100 kN in a direction parallel to the web.

Find the position and value of the maximum shearing stress, and the values of the shearing stresses at the top and bottom of the web; find also the amount of the shearing force taken by the web.

Prove briefly any formula used. (*UL*)

Answer 285 mm from top edge; 47.9 MN/m²; 22.6 and 33.8 MN/m²; 98.56 kN.

11 A cantilever, length L and of rectangular cross-section ($b \times d$), carries a load W. Derive expressions for the deflection at the free end due to shear if the load is (a) concentrated at the free end, (b) uniformly distributed throughout.

Answer (a) $6WL/5Gbd$; (b) $3WL/5Gbd$.

12 A beam of rectangular cross-section ($b \times d$) is simply supported over a span L. It carries a total load W uniformly distributed over the span. Obtain an expression for the deflection at mid-span due to shear.

If, in such a beam, the length is 20 times the depth and $E = 2.5G$ find the ratio of the deflection at mid-span due to shear to that due to bending.

Answer $3WL/20Gbd$; 3/500.

13 For a given cantilever of rectangular cross-section, length l, depth d and carrying a concentrated load at the free end, show that

$$\frac{\text{deflection due to shearing strain}}{\text{bending deflection}} = \text{constant} \times (d/l)^2$$

and find the value of the constant for a steel cantilever. Hence find the least value of l/d if the deflection due to shearing strain is not to exceed 1 per cent of the total deflection.

For steel $E = 203 \text{ GN/m}^2$, $G = 78 \text{ GN/m}^2$. (*UL*)

Answer 0.781; 8.836.

14 Derive an expression for the deflection due to shear in a cantilever of rectangular section, loaded with a concentrated load at the free end. The usual parabolic distribution of shearing stress over the cross-section may be assumed.

A cantilever of length L, rectangular in section, of depth d and breadth b, carries two point loads, each W, one at the free end and the other half-way along its length. If E/G is 2.5 find the ratio of d to L, for which the deflection at the end due to shear will be 1/100 of that due to bending. (UL)

Answer $6WL/5Gbd$; 0.108.

15 An I-beam has flanges 8 cm wide and 1 cm thick and a web $\frac{1}{2}$ cm thick. The overall depth is 18 cm.

When loaded as a beam there is, at a certain section, a bending moment M and a shearing force S.

At a point on this section 6 cm below the neutral layer, the strains are measured along the three directions OA, OB and OC as indicated in Figure 6.15.

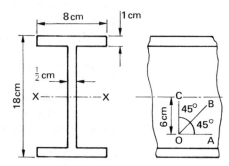

Figure 6.15

The values of these strains are as follows:

Along OA, $+3.42 \times 10^{-4}$; along OB, $+3.64 \times 10^{-4}$; along OC, -1.14×10^{-4}.

Determine

(a) the principal strains and stresses at O

(b) the value of the bending moment M and the shearing force S.

Take I_{xx} for the beam section 1328 cm⁴; $E = 208$ GN/m² and Poisson's ratio as 0.3. (UL)

Answer (a) $+4.52$ and -2.24×10^{-4}; $+88.0$ and -20.3 MN/m²;
(b) 13.67 kN m; 374 N.

16 Define the 'shear centre' of a beam.

A beam of channel section carries a vertical load and is supported so that the two flanges are horizontal. The flanges and web have equal thicknesses which are small compared to the depth of the web (D) and the width of the flanges (B).

Show by working from first principles that the shear centre is at a distance $3B^2/(6B + D)$ from the web.

Assumptions must be clearly stated, where they are made.

(UL)

17 A timber beam of uniform rectangular section has a length of 840 mm; the breadth of the beam is 30 mm and the depth is 70 mm. The beam is simply supported at its ends and it carries two loads 240 mm apart each W kN symmetrically situated at 300 mm from the ends of the beam.

Determine the magnitudes of the loads for a maximum stress due to bending of 7 MN/m² and find the maximum deflection of the beam due to both bending and shear. Formulae for deflection due to bending and shear must be derived, but in the case of shear the general formula for the distribution of shearing stress over the depth of the beam may be assumed.

For timber take $E = 10$ GN/m² and $G = 0.6$ GN/m². *(UL)*

Answer 1.143 kN; 2.93 mm; 0.33 mm.

18 The cantilever illustrated in Figure 6.16 is of solid circular section and is eccentrically loaded at the free end. At the fixed end A and B are at the ends of the vertical and horizontal diameters respectively.

Calculate the stresses at A and B due to bending moment, shear force and torsion. *Note*: The centroid of a semi-circular lamina is $4r/3\pi$ from the centre. *(PCL)*

Answer All in MN/m². At A: 293.4, 0 and 32.6; At B: 0, 2.72 and 32.6.

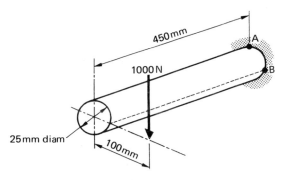

Figure 6.16

7

Laterally and eccentrically loaded struts

Standard cases

Table 7.1 shows the loading configurations and maximum bending moments for some common cases of slender struts pin-jointed at their ends. In each one P is the end load, L is the length, $\alpha^2 = P/EI$ and $P_E = \pi^2 EI/L^2$, the Euler load.

Table 7.1

Type of loading	Loading diagram	Maximum bending moment
End load having eccentricity e		$Pe \sec \frac{1}{2}\alpha L$
Strut initially bowed to sine wave with maximum deflection y_0. End load acting through centroid of end cross-section		$\dfrac{P_E}{P_E - P} Py_0$
Axial end load together with central lateral load W		$\dfrac{W}{2\alpha} \tan \frac{1}{2}\alpha L$
Axial end load together with uniformly distributed lateral load w per unit length		$\dfrac{w}{\alpha^2} (\sec \frac{1}{2}\alpha L - 1)$

Non-standard cases

Figure 7.1 shows a strut with an end load, point and distributed lateral loads and end fixing moments. With the notation given on the diagram and an origin at the left-hand support, the bending moment equation takes the form:

$$M = Py + R_1 x - M_1 + \text{term(s) due to lateral loads} \qquad (i)$$

In the absence of lateral load terms the combination of (i) and the differential equation of flexure gives

$$EI\frac{d^2y}{dx^2} = -M = -Py - R_1x + M_1 \tag{ii}$$

and the solution may be written

$$y = A\cos\alpha x + B\sin\alpha x - \frac{R_1x}{P} + \frac{M_1}{P} \tag{iii}$$

where A and B are integration constants. The determination of A, B, R_1 and M_1 depends on the end conditions. At a pinned or free end $M = 0$. If both ends are pinned R_1 (and R_2) can be determined at the outset by taking moments. If the ends are clamped it will be necessary to derive dy/dx from (iii) and use the condition $dy/dx = 0$ at the ends.

Lateral loads give rise to an extra term or terms in (ii), normally involving powers of x. The corresponding addition to (iii) will be of the form $a + bx + cx^2 \ldots$, the highest power of x being the same as that in (ii). The constants $a, b, c \ldots$ are then determined by differentiation of the result and comparison with the original differential equation.

In symmetrical cases it is often convenient to restrict the analysis to one-half of the strut and to use the conditions at mid-span for determining the integration constants.

In certain problems, where the ends of the struts are pin-jointed, the working can be simplified by differentiating equation (i) twice. This gives

$$\frac{d^2M}{dx^2} = P\frac{d^2y}{dx^2} + \text{term(s) which depend on the lateral loads}$$

and d^2y/dx^2 may be written $-M/EI$ from the differential equation of flexure. The result is an equation connecting M and x which may be solved by methods similar to those used previously for y. (See Examples 7.5 and 7.6.)

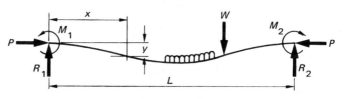

Figure 7.1

Worked examples

7.1 Maximum bending moment in strut with eccentric load

Obtain an expression for the maximum bending moment in a slender strut length L, pin-jointed at its ends and subjected to a compressive load P which acts parallel to the axis of the undeflected strut but at an eccentricity e. Show how the result leads to the formula for the Euler critical load.

A tubular strut length 1.5 m and diameter 50 mm has a wall thickness of 1 mm. An axial load of 10 kN is applied at the ends of the strut (which may be assumed as pin-jointed) and due to the load being slightly eccentric it is found that a central deflection is produced of magnitude 3 mm. Taking the eccentricity as being the same at each end, find its magnitude. $E = 70 \text{ GN/m}^2$.

Solution In the notation of Figure 7.2 the bending moment at distance x from one end is

$$M = P(y + e)$$

and the differential equation of flexure becomes

$$EI\frac{d^2y}{dx^2} = -M = -P(y + e)$$

or

$$\frac{d^2y}{dx^2} + \alpha^2(y + e) = 0$$

where $\alpha^2 = P/EI$.

The solution may be written

$$y + e = A \cos \alpha x + B \sin \alpha x$$

where A and B are constants.

When $x = 0$, $y = 0$ and therefore $A = e$. Thus

$$y + e = e \cos \alpha x + B \sin \alpha x \qquad \text{(i)}$$

When $x = L$, $y = 0$ and

$$0 + e = e \cos \alpha L + B \sin \alpha L$$

from which

$$B = \frac{e(1 - \cos \alpha L)}{\sin \alpha L} \qquad \text{(ii)}$$

It is convenient to use the half-angle $\frac{1}{2}\alpha L$ and, with the relationships $\cos \alpha L = 2 \cos^2 \frac{1}{2}\alpha L - 1$ and $\sin \alpha L = 2 \sin \frac{1}{2}\alpha L \cos \frac{1}{2}\alpha L$, we obtain

$$B = e \tan \tfrac{1}{2}\alpha L$$

(This result can also be obtained from the condition $dy/dx = 0$ at mid-span.)

Figure 7.2

With this value of B and using (i), the deflection at mid-span ($x = L/2$) is given by:

$$y_{max} + e = e \cos \tfrac{1}{2}\alpha L + e \tan \tfrac{1}{2}\alpha L \sin \tfrac{1}{2}\alpha L$$

$$= e\left(\frac{\cos^2 \tfrac{1}{2}\alpha L + \sin^2 \tfrac{1}{2}\alpha L}{\cos \tfrac{1}{2}\alpha L}\right)$$

$$= e \sec \tfrac{1}{2}\alpha L$$

so that

$$y_{max} = e(\sec \tfrac{1}{2}\alpha L - 1) \qquad\qquad\qquad\qquad \text{(iii)}$$

The maximum bending moment is therefore

$$M_{max} = P(y_{max} + e) = Pe \sec \tfrac{1}{2}\alpha L \qquad\qquad (Ans)$$

The value of $\sec \theta$ becomes infinite when $\theta = \pi/2$. Thus, if $\tfrac{1}{2}\alpha L = \pi/2$ the bending moment is (theoretically) infinite and the strut will collapse however small the eccentricity of the load. The corresponding value of P is given by

$$\alpha L = \pi \quad \text{or} \quad \alpha^2 L^2 = \pi^2$$

But $\alpha^2 = P/EI$ and hence the critical value of P is

$$P = \pi^2 EI/L^2$$

the Euler expression.

In the numerical part of the question let 50 mm be the internal diameter. Then

$$I = \frac{\pi}{64}(52^4 - 50^4) = 5.213 \times 10^4 \text{ mm}^4$$

$$\alpha^2 = \frac{P}{EI} = \frac{(10 \times 10^3 \text{ N})}{(70 \times 10^9 \text{ N/m}^2) \times (5.213 \times 10^4 \times 10^{-12} \text{ m}^4)}$$

$$= 2.742 \text{ m}^{-2}$$

$$\alpha = 1.656 \text{ m}^{-1}$$

From (iii) the eccentricity of the load is

$$e = \frac{y_{max}}{\sec \tfrac{1}{2}\alpha L - 1} = \frac{3 \text{ mm}}{\sec(\tfrac{1}{2} \times 1.656 \times 1.5) - 1}$$

$$= 1.32 \text{ mm} \qquad\qquad\qquad\qquad\qquad (Ans)$$

7.2 Maximum stress in initially bowed strut

A hollow circular steel strut with its ends position-fixed has a length of 2.5 m, its external diameter being 100 mm and its internal diameter 80 mm. Before loading the strut is bent with a maximum deviation of 4 mm. Assuming that the centre line of the strut is sinusoidal determine the maximum stress due to a central compressive end load of 100 kN. ($E = 206 \text{ GN/m}^2$.)

(UL)

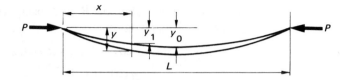

Figure 7.3

Solution Figure 7.3 shows a strut, length L, subjected to an end compressive load P. Suppose that at distance x from the left-hand end, y_1 is the deflection due to the initial bending and y is the total deflection when the load is applied. Let y_0 be the maximum initial deviation. Since the initial bowing is sinusoidal we have

$$y_1 = C \sin Dx$$

where C and D are constants.

When $x = L$, $y_1 = 0$ and thus $\sin DL = 0$. The lowest value which satisfies the problem is $DL = \pi$ and hence $D = \pi/L$.

At mid-span $x = L/2$ and $y_1 = y_0$. Using the value for D found above we obtain

$$y_0 = C \sin \left(\frac{\pi}{L} \times \frac{L}{2} \right)$$

from which $C = y_0$. Thus the initial deflection is given by

$$y_1 = y_0 \sin \frac{\pi x}{L} \tag{i}$$

At distance x from the origin the bending moment due to P is Py but the change in deflection due to the load is $y - y_1$. Hence the differential equation of flexure becomes

$$EI \frac{d^2(y - y_1)}{dx^2} = -M = -Py$$

or

$$\frac{d^2y}{dx^2} - \frac{d^2y_1}{dx^2} = -\frac{P}{EI} y = -\alpha^2 y \tag{ii}$$

From (i),

$$\frac{d^2y_1}{dx^2} = -\frac{\pi^2 y_0}{L^2} \sin \frac{\pi x}{L}$$

and (ii) becomes

$$\frac{d^2y}{dx^2} = -\alpha^2 y - \frac{\pi^2 y_0}{L^2} \sin \frac{\pi x}{L}$$

The solution of this equation can be written

$$y = A \cos \alpha x + B \sin \alpha x + \frac{\pi^2 y_0}{\pi^2 - \alpha^2 L^2} \sin \frac{\pi x}{L} \qquad \text{(iii)}$$

and this result can be checked by successive differentiation.
When $x = 0$, $y = 0$ and therefore $A = 0$.
When $x = L$, $y = 0$ and thus $B \sin \alpha L = 0$.
For $\sin \alpha L$ to be zero, $\alpha L = \pi$ (the lowest value satisfying the problem) but this is ruled out because it would make the third term of (iii) infinite. Hence B must be zero and equation (iii) becomes

$$y = \frac{\pi^2 y_0}{\pi^2 - \alpha^2 L^2} \sin \frac{\pi x}{L}$$

The greatest deflection occurs at mid-span ($x = L/2$) and thus

$$y_{max} = \frac{\pi^2 y_0}{\pi^2 - \alpha^2 L^2} = \frac{y_0}{1 - (\alpha^2 L^2 / \pi^2)}$$

But $\alpha^2 = P/EI$ and $\alpha^2 L^2 / \pi^2 = P/P_E$ where P_E is the Euler load for the strut. Thus

$$y_{max} = \left(\frac{1}{1 - (P/P_E)} \right) y_0 = \left(\frac{P_E}{P_E - P} \right) y_0$$

For the given steel strut,

$$I = \frac{\pi}{64} (100^4 - 80^4) \, \text{mm}^4 = 2.90 \times 10^6 \, \text{mm}^4$$

$$P_E = \frac{\pi^2 EI}{L^2} = \frac{\pi^2 \times (206 \times 10^9 \, \text{N/m}^2) \times (2.90 \times 10^{-6} \, \text{m}^4)}{(2.5 \, \text{m})^2}$$

$$= 942 \, \text{kN}$$

$$y_{max} = \left(\frac{P_E}{P_E - P} \right) y_0 = \left(\frac{942}{942 - 100} \right) \times 4 \, \text{mm}$$

$$= 4.47 \, \text{mm}$$

The maximum bending moment is

$$P y_{max} = 100 \, \text{kN} \times 0.004 \, 47 \, \text{m} = 447 \, \text{N m}$$

and the corresponding maximum bending stress is

$$\sigma = \frac{447 \, \text{N m} \times 0.05 \, \text{m}}{2.90 \times 10^{-6} \, \text{m}^4} = 7.71 \, \text{MN/m}^2$$

$$\text{Axial compressive stress} = \frac{0.1 \, \text{MN}}{\frac{1}{4}\pi(100^2 - 80^2) \times 10^{-6} \, \text{m}^2}$$

$$= 35.38 \, \text{MN/m}^2$$

Maximum stress $= 7.71 + 35.38 = 43.1 \, \text{MN/m}^2$ (compressive)

(*Ans*)

7.3 Perry–Robertson formula for maximum stress

What are the basic assumptions made in deriving the Perry–Robertson formula for struts?

For what reasons was this formula made the basis of practical design formulae?

Show that if a strut has an initial curvature in the form of a parabolic arc and is hinged at both ends (i.e. position fixed only) then the maximum compressive stress produced by a load P is

$$\sigma_{max} = \frac{P}{A}\left\{1 + \frac{es}{k^2}\frac{8P_E}{\pi^2 P}\left(\sec \tfrac{1}{2}\pi \sqrt{\left[\frac{P}{P_E}\right]} - 1\right)\right\}$$

where A is the cross-sectional area; e the initial central deflection; P_E the Eulerian crippling load; k the least radius of gyration of the section and s the distance of the extreme fibres on the compression side from the neutral axis.

Compare the maximum stress *due to bending only* derived from the given formula with that given by the Perry–Robertson assumptions when $P = \tfrac{1}{2}P_E$. (UL)

Solution The Perry–Robertson formula for pin-jointed struts with end loads is based on the following assumptions:

(a) the strut is initially bowed to a sine wave with a maximum deviation at the centre (say e_1);
(b) the load is applied eccentrically (by an amount, say, e_2);
(c) the strut fails when the maximum compressive stress reaches the yield stress of the material in direct compression (say σ_Y).

As shown in the solution to Example 7.2 assumption (a) leads to a maximum deflection of:

$$y_{max} = \frac{P_E}{P_E - P}e_1 \tag{i}$$

Similarly the analysis given in Example 8.3 of Volume 1 shows that an initial eccentricity e_2 leads to a maximum deflection

$$y_{max} = e_2(\sec \tfrac{1}{2}\alpha L - 1)$$

and the corresponding moment arm of the load P is

$$y_{max} + e_2 = e_2 \sec \tfrac{1}{2}\alpha L = e_2 \sec \tfrac{1}{2}\pi\sqrt{(P/P_E)} \tag{ii}$$

Perry showed that the expression $\sec \tfrac{1}{2}\pi\sqrt{(P/P_E)}$ is very nearly equal to $1.2P_E/(P_E - P)$ and thus (ii) becomes

$$y_{max} + e_2 = \left(\frac{P_E}{P_E - P}\right)1.2e_2 \tag{iii}$$

A comparison of (i) and (iii) shows that the separate effects of e_1 and e_2 can be simultaneously allowed for by an equivalent initial bowing e

given by

$$e = e_1 + 1.2e_2$$

With the notation of the present question and assumption (c) above the maximum compressive stress is given by

$$\sigma_{max} = \text{direct stress} + \text{maximum compressive bending stress}$$

$$= \frac{P}{A} + \frac{Pes}{I}\left(\frac{P_E}{P_E - P}\right)$$

$$= \sigma + \frac{\sigma es}{k^2}\left(\frac{\sigma_E}{\sigma_E - \sigma}\right)$$

where $\sigma = P/A$ and $\sigma_E = \text{Euler stress} = P_E/A$.

On rearrangement, the last equation becomes a quadratic

$$\sigma^2 - \sigma\left[\sigma_{max} + \sigma_E\left(1 + \frac{es}{k^2}\right)\right] + \sigma_E\sigma_{max} = 0$$

from which, taking the smaller root,

$$\sigma = \tfrac{1}{2}[\sigma_{max} + (1 + \eta)\sigma_E] - \tfrac{1}{2}\surd\{[\sigma_{max} + (1 + \eta)\sigma_E]^2 - 4\sigma_E\sigma_{max}\}$$

where $\eta = es/k^2$.

This result is known as the Perry–Robertson formula and is the basis for the calculation of safe loads for steel columns (see British Standard 449: Part 1 1970 and Part 2 1969).

This formula has been adopted for practical design because there is inevitably some initial crookedness and eccentricity of loading. The expression for σ represents closely the results obtained from experiments in which these imperfections are kept as small as possible. Robertson analysed a considerable number of results and concluded that for structural materials with a marked yield point η could be expressed as a simple function of the slenderness ratio L/k.

Figure 7.3 illustrates an initially bowed strut, the only difference in the present case being that the maximum initial deviation is e instead of y_0. Since the initial shape is now parabolic the deflection before loading is given by

$$y_1 = a + bx + cx^2 \tag{iv}$$

where a, b and c are constants. When $x = 0$, $y_1 = 0$ and $a = 0$. At $x = \tfrac{1}{2}L$ the slope $dy_1/dx = 0$ and hence $b = -cL$.

In addition, when $x = \tfrac{1}{2}L$, $y_1 = e$. This gives $b = 4e/L$ and $c = -4e/L^2$. Thus, from (iv),

$$y_1 = 4e\left(\frac{x}{L} - \frac{x^2}{L^2}\right) \tag{v}$$

The differential equation of flexure is now

$$EI\frac{d^2(y - y_1)}{dx^2} = -M = -Py$$

or

$$\frac{d^2y}{dx^2} - \frac{d^2y_1}{dx^2} = -\alpha^2 y \qquad \text{(vi)}$$

From (v),

$$\frac{d^2y_1}{dx^2} = -\frac{8e}{L^2}$$

and equation (vi) becomes

$$\frac{d^2y}{dx^2} = -\alpha^2 y - \frac{8e}{L^2}$$

The solution of this equation is

$$y = A\cos\alpha x + B\sin\alpha x - \frac{8e}{\alpha^2 L^2} \qquad \text{(vii)}$$

When $x = 0$, $y = 0$ and $A = 8e/\alpha^2 L^2$.
When $x = L$, $y = 0$ and

$$B = \frac{8e}{\alpha^2 L^2}\operatorname{cosec}\alpha L(1 - \cos\alpha L)$$

But $\operatorname{cosec}\alpha L$ and $\cos\alpha L$ can be written in terms of the half-angles $\frac{1}{2}\alpha L$ so that

$$B = \frac{8e}{\alpha^2 L^2}\frac{1}{2\sin\frac{1}{2}\alpha L\cos\frac{1}{2}\alpha L}(2\sin^2\tfrac{1}{2}\alpha L)$$

$$= \frac{8e}{\alpha^2 L^2}\tan\tfrac{1}{2}\alpha L$$

The maximum deflection occurs at mid-span ($x = L/2$) and substituting for A and B, equation (vii) gives

$$y_{max} = \frac{8e}{\alpha^2 L^2}\cos\tfrac{1}{2}\alpha L + \frac{8e}{\alpha^2 L^2}\tan\tfrac{1}{2}\alpha L\sin\tfrac{1}{2}\alpha L - \frac{8e}{\alpha^2 L^2}$$

$$= \frac{8e}{\alpha^2 L^2}\left(\frac{\cos^2\frac{1}{2}\alpha L + \sin^2\frac{1}{2}\alpha L}{\cos\frac{1}{2}\alpha L} - 1\right)$$

$$= \frac{8e}{\alpha^2 L^2}(\sec\tfrac{1}{2}\alpha L - 1)$$

$$= \frac{8e}{\pi^2}\frac{P_E}{P}(\sec\tfrac{1}{2}\pi\sqrt{(P/P_E)} - 1)$$

since

$$\alpha^2 = \frac{P}{EI} = \frac{\pi^2}{L^2}\frac{P}{P_E}$$

The maximum bending moment is Py_{max} and the maximum compressive stress is therefore:

direct stress + maximum bending stress

$$= \frac{P}{A} + \frac{M_{max}s}{I}$$

$$= \frac{P}{A} + \frac{Ps}{I} \frac{8e}{\pi^2} \frac{P_E}{P} \left(\sec \tfrac{1}{2}\pi \sqrt{\left[\frac{P}{P_E}\right]} - 1 \right)$$

$$= \frac{P}{A} \left\{ 1 + \frac{es}{k^2} \frac{8P_E}{\pi^2 P} \left(\sec \tfrac{1}{2}\pi \sqrt{\left[\frac{P}{P_E}\right]} - 1 \right) \right\} \qquad (Ans)$$

since $I = Ak^2$.

If $P = \tfrac{1}{2}P_E$ the maximum bending moment will be

$$Py_{max} = \frac{8e}{\pi^2} \times 2P(\sec \tfrac{1}{2}\pi\sqrt{\tfrac{1}{2}} - 1) = 2.03eP \qquad (viii)$$

With the basic Perry–Robertson assumption of sinusoidal bowing we have

$$M_{max} = \left(\frac{P_E}{P_E - P}\right)eP = \left(\frac{1}{1 - \tfrac{1}{2}}\right)eP = 2eP \qquad (ix)$$

Comparing (viii) and (ix) we see that the formula in the question gives a slightly higher result than the Perry–Robertson assumptions.

7.4 Maximum deflection of strut with axial end load and central lateral point load

A slender strut of uniform section and of length L has pin-jointed ends and it is initially straight and vertical. It carries an axial load P and also a horizontal lateral load W applied at the middle of its length and acting in the plane in which P would cause bending to occur. Show that the maximum deflection is

$$\Delta = \frac{W}{2\alpha P} \tan \tfrac{1}{2}\alpha L - \frac{WL}{4P}$$

where

$$\alpha = \sqrt{(P/EI)}$$

In the case of a given strut the magnitude of P is $P = P_E/4$ where P_E is the Euler critical load for the strut. Find the ratio of the maximum deflection produced by P and the lateral load W acting together, to that produced by W acting alone. (UL)

Solution Figure 7.4 gives the notation, the strut being shown horizontal for ease of comparison with other examples. By symmetry the lateral reactions at the ends are each $W/2$. The differential

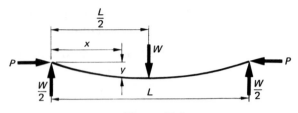

Figure 7.4

equation of flexure is given by

$$EI\frac{d^2y}{dx^2} = -M = -Py - \tfrac{1}{2}Wx$$

for the range $0 \leqslant x \leqslant L/2$.

The solution of this equation is

$$y = A \cos \alpha x + B \sin \alpha x - Wx/2P \qquad\qquad\text{(i)}$$

where $\alpha^2 = P/EI$ and A and B are constants.

When $x = 0$, $y = 0$ and thus, from (i), $A = 0$.

By differentiation the slope is

$$dy/dx = \alpha B \cos \alpha x - W/2P \qquad\qquad\text{(ii)}$$

Since equations (i) and (ii) apply only to the left-hand half of the strut the constant B must be determined from the zero slope condition at mid-span.

Thus $dy/dx = 0$ when $x = L/2$ and

$$\alpha B \cos \tfrac{1}{2}\alpha L = W/2P$$

from which

$$B = \frac{W}{2\alpha P}\sec \tfrac{1}{2}\alpha L$$

Using the values obtained for A and B, the maximum deflection, which occurs at mid-span, is given by (i) with $x = L/2$. Hence

$$\Delta \text{ (or } y_{\text{max}}) = \frac{W}{2\alpha P}\sec \tfrac{1}{2}\alpha L \sin \tfrac{1}{2}\alpha L - \frac{W}{2P}\frac{L}{2}$$

$$= \frac{W}{2\alpha P}\tan \tfrac{1}{2}\alpha L - \frac{WL}{4P}$$

as required. Although the question does not call for the maximum bending moment it occurs at mid-span and is easily found from this last result to be

$$M_{\text{max}} = \frac{W}{2\alpha}\tan \tfrac{1}{2}\alpha L$$

The Euler load $P_E = \pi^2 EI/L^2$ and hence $P = \pi^2 EI/4L^2$ in the present example.

Also

$$\alpha^2 = P/EI = \pi^2/4L^2 \quad \text{and} \quad \alpha = \pi/2L$$

The maximum deflection with P and W acting together is therefore

$$\Delta = \frac{W}{2\alpha P} \tan \tfrac{1}{2}\alpha L - \frac{WL}{4P}$$

$$= \frac{W}{2(\pi/2L)(\pi^2 EI/4L^2)} \tan \frac{\pi}{2L}\frac{L}{2} - \frac{WL}{4(\pi^2 EI/4L^2)}$$

$$= \frac{4WL^3}{\pi^3 EI} \tan \tfrac{1}{4}\pi - \frac{WL^3}{\pi^2 EI}$$

$$= \frac{WL^3}{EI} \left(\frac{4}{\pi^3} - \frac{1}{\pi^2} \right)$$

The maximum deflection with W acting alone is

$$\Delta = WL^3/48EI$$

The required ratio is therefore

$$\left(\frac{4}{\pi^3} - \frac{1}{\pi^2} \right) \Big/ \frac{1}{48}$$

which equals 1.33.

7.5 Maximum stress in strut with axial end load and uniformly distributed lateral load

Show that

$$M = \frac{wEI}{P} \left(\sec \tfrac{1}{2}l \sqrt{\left[\frac{P}{EI} \right]} - 1 \right)$$

represents the maximum bending moment in a horizontal strut of length l, having pin-jointed ends, if it weighs w per unit length and is subjected to an axial load P.

A horizontal pin-ended strut 5 m long is formed from a standard rolled steel Tee section (150 mm × 100 mm × 12.5 mm @ 24.15 kg/m). The axial compressive load is 100 kN. Determine the maximum stress in the steel if the XX axis is horizontal and the table of the Tee forms the compression face. The centroid of the section is 2.5 cm below the top. ($I_{XX} = 253$ cm⁴; cross-sectional area = 30.8 cm²).

$$E = 206 \text{ GN/m}^2. \tag{UL}$$

Solution The weight of the strut is equivalent to a uniformly distributed load and Figure 7.5 shows the effective loading. The bending moment at a distance x from one end is

$$M = Py + \tfrac{1}{2}wlx - \tfrac{1}{2}wx^2 \tag{i}$$

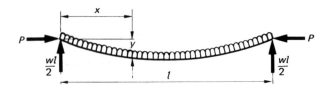

Figure 7.5

An expression for the deflection is not required and the following method enables us to determine M in terms of x without first finding the deflection equation.

Differentiating equation (i) twice with respect to x,

$$\frac{d^2M}{dx^2} = P\frac{d^2y}{dx^2} - w = -\frac{P}{EI}M - w$$

since $EI(d^2y/dx^2) = -M$.

The solution of this equation is

$$M = A\cos\alpha x + B\sin\alpha x - w/\alpha^2 \tag{ii}$$

where $\alpha^2 = P/EI$.

The strut is pin-jointed at the ends and, when $x = 0$, $M = 0$. From (i), therefore, $A = w/\alpha^2$.

With this value and the condition $x = l$, $M = 0$, we have

$$\frac{w}{\alpha^2}\cos\alpha l + B\sin\alpha l = \frac{w}{\alpha^2}$$

from which

$$B = \frac{w}{\alpha^2}\frac{(1-\cos\alpha l)}{\sin\alpha l}$$

Using the trigonometrical identities $\cos\alpha l = 1 - 2\sin^2\frac{1}{2}\alpha l$ and $\sin\alpha l = 2\sin\frac{1}{2}\alpha l\cos\frac{1}{2}\alpha l$ this becomes

$$B = \frac{w}{\alpha^2}\tan\frac{1}{2}\alpha l$$

The maximum bending moment occurs at mid-span where $x = l/2$ and with the values obtained for A and B, equation (ii) gives

$$M_{\text{max}} = \frac{w}{\alpha^2}\cos\frac{1}{2}\alpha l + \frac{w}{\alpha^2}\tan\frac{1}{2}\alpha l\sin\frac{1}{2}\alpha l - \frac{w}{\alpha^2}$$

$$= \frac{w}{\alpha^2}\left(\frac{\cos^2\frac{1}{2}\alpha l + \sin^2\frac{1}{2}\alpha l}{\cos\frac{1}{2}\alpha l} - 1\right)$$

$$= \frac{w}{\alpha^2}(\sec\frac{1}{2}\alpha l - 1)$$

This is the required result since $\alpha^2 = P/EI$.

For the given Tee section,

$$\text{axial compressive stress} = \frac{100 \times 10^3\,\text{N}}{30.8 \times 10^{-4}\,\text{m}^2}$$

$$= 32.5\,\text{MN/m}^2$$

$$\alpha^2 = \frac{P}{EI} = \frac{(100 \times 10^3)\,\text{N}}{(206 \times 10^9)\,\text{N/m}^2 \times (253 \times 10^{-8})\,\text{m}^4}$$

$$= 0.191\,9\,\text{m}^{-2} \qquad \alpha = 0.438\,\text{m}^{-1}$$

The reference '@ 24.15 kg/m' means that the strut weighs 24.15 kgf per metre and thus $w = 24.15 \times 9.81 = 237\,\text{N/m}$.

Also $l = 5\,\text{m}$ and the maximum bending moment is

$$M_{max} = \frac{w}{\alpha^2}(\sec \tfrac{1}{2}\alpha l - 1)$$

$$= \frac{237\,\text{N/m}}{0.191\,9\,\text{m}^{-2}}[\sec(\tfrac{1}{2} \times 0.438 \times 5) - 1]$$

$$= 1\,461\,\text{N m}$$

The maximum compressive stress due to bending occurs at the top edge (see Figure 7.6). If y_{max} is now the distance from the centroid of the section to this edge, maximum compressive bending stress is

$$\frac{M y_{max}}{I} = \frac{(1.461 \times 10^3\,\text{N m}) \times (25 \times 10^{-3}\,\text{m})}{253 \times 10^{-8}\,\text{m}^4}$$

$$= 14.4\,\text{MN/m}^2$$

Combining this with the axial compressive stress gives a maximum compressive stress of

$$32.5 + 14.4 = 46.9\,\text{MN/m}^2 \qquad\qquad\qquad (Ans)$$

At the bottom edge of the section, $y_{max} = 7.5\,\text{cm}$, the maximum tensile stress is found to be $43.2\,\text{MN/m}^2$ and the resultant of this and the axial compressive stress is $10.7\,\text{MN/m}^2$ (tensile).

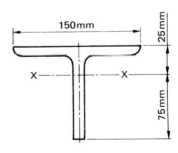

Figure 7.6

7.6 Maximum bending moment in strut with axial end load and transverse distributed load of varying intensity

A straight strut of length L and of uniform section is hinged at both ends and is loaded along its axis with a thrust P. It also carries a transverse distributed load which varies uniformly in intensity from w per unit length at one end A to zero at the other end B.

Show that the distance x from the end B to the section at which the maximum bending moment occurs is given by

$$\cos \alpha x = \frac{\sin \alpha L}{\alpha L}$$

where $\alpha^2 = P/EI$.

If the thrust P is 81 per cent of the Eulerian crippling load, find the position and value of the maximum bending moment. (UL)

Solution Take the origin at the end B. Then, with the notation of Figure 7.7 the intensity of loading at a section XX is wx/L, the average intensity to the left of XX is $wx/2L$, the total lateral load on the portion B to XX is $wx^2/2L$ and its moment arm about XX is $x/3$.

The bending moment equation is therefore

$$M = Py + \frac{wLx}{6} - \frac{wx^3}{6L}$$

and, by differentiating twice with respect to x,

$$\frac{d^2M}{dx^2} = P\frac{d^2y}{dx^2} - \frac{wx}{L} = -\alpha^2 M - \frac{wx}{L}$$

(since $d^2y/dx^2 = -M/EI$).

The solution of this equation is

$$M = A \cos \alpha x + B \sin \alpha x - wx/\alpha^2 L \qquad (i)$$

When $x = 0$, $M = 0$ (pin-jointed) and thus $A = 0$.
Also when $x = L$, $M = 0$ and $B = (w/\alpha^2) \operatorname{cosec} \alpha L$.
For maximum (or minimum) $dM/dx = 0$ and this gives

$$\alpha B \cos \alpha x - \frac{w}{\alpha^2 L} = 0$$

Using the value obtained above for B we have

$$\alpha \left(\frac{w}{\alpha^2} \operatorname{cosec} \alpha L \right) \cos \alpha x = \frac{w}{\alpha^2 L}$$

$$\cos \alpha x = \frac{\sin \alpha L}{\alpha L}$$

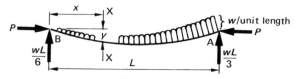

Figure 7.7

as required. From physical considerations this is a maximum rather than a minimum.

The Euler load is $\pi^2 EI/L^2$ and hence

$$P = 0.81\pi^2 EI/L^2$$

Therefore

$$\alpha^2 = 0.81\pi^2/L^2 \quad \text{and} \quad \alpha = 0.9\pi/L$$

The position of maximum bending moment is given by

$$\cos \alpha x = (\sin \alpha L)/\alpha L = (\sin 0.9\pi)/0.9\pi = 0.109$$

$$\alpha x = 83.7° \quad \text{or} \quad 1.46 \text{ rad}$$

and the position of the maximum bending moment is given by

$$x = \frac{1.46}{0.9\pi/L} = 0.517L \qquad (Ans)$$

With $\alpha = 0.9\pi/L$ and $x = 0.517L$ the maximum bending moment can be found from (i). It is

$$
\begin{aligned}
M_{\max} &= \frac{w}{\alpha^2} \operatorname{cosec} \alpha L \sin \alpha x - \frac{wx}{\alpha^2 L} \\
&= \frac{wL^2}{0.81\pi^2} \operatorname{cosec}\left(\frac{0.9\pi}{L}L\right) \sin\left(\frac{0.9\pi}{L} \times 0.517L\right) \\
&\quad - \frac{w \times 0.517L \times L^2}{0.81\pi^2 \times L} \\
&= \frac{wL^2}{0.81\pi^2} (\operatorname{cosec} 162° \sin 83.7° - 0.517) \\
&= 0.338wL^2 \qquad (Ans)
\end{aligned}
$$

7.7 Bending moments at ends and centre of strut with built-in ends, axial end load and uniformly distributed lateral load

Obtain expressions for the bending moments at the ends and centre of a uniform strut, built in at both ends, and subjected to a uniform lateral load of intensity w. The strut length is L, the end thrust P, and the elastic properties EI. Take $\alpha^2 = P/EI$. Show, without elaborate analysis, from the expressions derived that for practical struts the end moments are greater numerically than the central moment and of opposite sign. (UL)

Solution Figure 7.8 illustrates the notation. By symmetry the end reactions $R_1 = R_2 = wL/2$ and $M_1 = M_2$. The differential equation of flexure gives

$$EI\frac{d^2y}{dx^2} = -M = -Py - R_1 x + M_1 + \tfrac{1}{2}wx^2 \qquad (i)$$

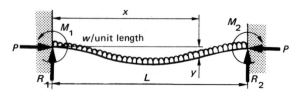

Figure 7.8

and the solution may be written

$$y = A \cos \alpha x + B \sin \alpha x + (a + bx + cx^2) \qquad \text{(ii)}$$

where the constants a, b and c can be determined by substituting (ii) in (i) and equating coefficients of powers of x.

From (ii),

$$\frac{d^2y}{dx^2} = -\alpha^2(A \cos \alpha x + B \sin \alpha x) + 2c$$

$$= -\alpha^2(y - a - bx - cx^2) + 2c$$

and

$$EI\frac{d^2y}{dx^2} = -P(y - a - bx - cx^2) + 2cEI$$

On comparing coefficients of x in this expression with those in (i) we find

for x^2: $cP = w/2$ and $c = w/2P$

for x: $bP = -R_1$ and $b = -R_1/P$

for number term: $Pa + 2cEI = M_1$

$$a = \frac{M_1 - 2cEI}{P} = \frac{M_1}{P} - \frac{w}{\alpha^2 P}$$

(since $c = w/2P$).

Thus the deflection equation (ii) becomes

$$y = A \cos \alpha x + B \sin \alpha x + \frac{M_1}{P} - \frac{R_1 x}{P} + \frac{wx^2}{2P} - \frac{w}{\alpha^2 P} \qquad \text{(iii)}$$

When $x = 0$, $y = 0$ and from (iii)

$$A = \frac{w}{\alpha^2 P} - \frac{M_1}{P}$$

By differentiating (iii) the slope is given by

$$\frac{dy}{dx} = -\alpha A \sin \alpha x + \alpha B \cos \alpha x - \frac{R_1}{P} + \frac{wx}{P} \qquad \text{(iv)}$$

At $x = 0$, $dy/dx = 0$ and $B = R_1/\alpha P$.

To find M_1 it is necessary to use a further 'end' condition. It is convenient to use $dy/dx = 0$ when $x = L/2$ (from considerations of symmetry), and, from (iv) we have, noting that $R_1 = wL/2$,

$$0 = -\alpha A \sin \tfrac{1}{2}\alpha L + \alpha B \cos \tfrac{1}{2}\alpha L - \frac{wL}{2P} + \frac{wL}{2P}$$

$$\cot \tfrac{1}{2}\alpha L = \frac{A}{B} = \frac{2\alpha}{wL}\left(\frac{w}{\alpha^2} - M_1\right)$$

The end moment at $x = 0$ is therefore

$$M_1 = \frac{w}{\alpha^2} - \frac{wL}{2\alpha} \cot \tfrac{1}{2}\alpha L$$

and with this result the constant A is

$$\frac{wL}{2\alpha P} \cos \tfrac{1}{2}\alpha L$$

Combining (i) and (iii) and substituting for A, B, M_1 and R_1 the expression for bending moment becomes

$$M = Py + R_1 x - M_1 - \tfrac{1}{2}wx^2$$

$$= P\left(A \cos \alpha x + B \sin \alpha x - \frac{w}{\alpha^2 P}\right)$$

$$= \frac{wL}{2\alpha}(\cot \tfrac{1}{2}\alpha L \cos \alpha x + \sin \alpha x) - \frac{w}{\alpha^2}$$

The end moment M_0 is obtained by putting $x = 0$ and thus

$$M_0 = \frac{wL}{2\alpha} \cot \tfrac{1}{2}\alpha L - \frac{w}{\alpha^2} = \frac{wL}{2\alpha}\left(\cot \tfrac{1}{2}\alpha L - \frac{2}{\alpha L}\right)$$

Similarly the bending moment M_c at mid-span ($x = L/2$) is

$$M_c = \frac{wL}{2\alpha}\left(\frac{\cos \tfrac{1}{2}\alpha L}{\sin \tfrac{1}{2}\alpha L} \cos \tfrac{1}{2}\alpha L + \sin \tfrac{1}{2}\alpha L\right) - \frac{w}{\alpha^2}$$

$$= \frac{wL}{2\alpha} \operatorname{cosec} \tfrac{1}{2}\alpha L - \frac{w}{\alpha^2} = \frac{wL}{2\alpha}\left(\operatorname{cosec} \tfrac{1}{2}\alpha L - \frac{2}{\alpha L}\right)$$

For practical struts $\tfrac{1}{2}\alpha L$ must be less than $\tfrac{1}{2}\pi$. Denoting $\tfrac{1}{2}\alpha L$ by θ the two expressions become

$$M_0 = \frac{wL}{2\alpha}\left(\frac{1}{\tan \theta} - \frac{1}{\theta}\right)$$

and

$$M_c = \frac{wL}{2\alpha}\left(\frac{1}{\sin \theta} - \frac{1}{\theta}\right)$$

For practical values of θ, $\sin \theta < \theta < \tan \theta$ so that M_0 is negative and M_c positive. In addition, a consideration of corresponding values of θ, $\sin \theta$ and $\tan \theta$ shows that M_0 is numerically greater than M_c.

Problems

1 Show that the maximum bending moment M_{max} for a strut, pin-jointed at its ends, with an end load which is parallel to the undeflected line of the strut but eccentric to it is given by

$$\frac{M_{max}}{M_0} = \sec \tfrac{1}{2}\pi\sqrt{(P/P_E)}$$

where M_0 is the bending moment in the strut if its deflection is ignored and P_E = Euler load.

Plot M_{max}/M_0 against P/P_E for values of P/P_E from 0 to 0.81. (It is convenient to take intermediate values of 0.09, 0.16, 0.25, 0.36, 0.49 and 0.64.)

Answer For the suggested values of P/P_E the values of M_{max}/M_0 are: 1.00, 1.12, 1.24, 1.41, 1.70, 2.20, 3.24, 6.39.

2 Show, with the notation of the previous example, that if the end load is axial but the strut has an initial sinusoidal bowing the maximum bending moment is given by

$$\frac{M_{max}}{M_0} = \frac{1}{1 - P/P_E}$$

where M_0 is now the maximum bending moment due to the initial bowing (ignoring the further deflection due to bending).

Plot M_{max}/M_0 against P/P_E for the same range of values as before.

Answer 1.00, 1.10, 1.19, 1.33, 1.56, 1.96, 2.78, 5.26.

3 If, in Problem 1, the strut is straight and the end load is axial but there is a central lateral point load W, show that

$$M_{max}/M_0 = \tan\theta/\theta$$

where M_0 is now the maximum bending moment when $P = 0$ (i.e. $M_0 = WL/4$) and $\theta = \tfrac{1}{2}\pi\sqrt{(P/P_E)}$.

Plot M_{max}/M_0 against P/P_E for the same range of values as before.

If, however, the lateral load is uniformly distributed over the length of the strut, show that

$$\frac{M_{max}}{M_0} = \frac{2}{\theta^2}(\sec\theta - 1)$$

Plot M_{max}/M_0 against P/P_E on the same axes as for the central point load case. (*Note*: $M_0 = WL/8$, and, for small angles $\sec\theta \simeq 1 + \tfrac{1}{2}\theta^2$.)

Answer

P/P_E	0	0.09	0.16	0.25	0.36	0.49	0.64	0.81
$\dfrac{M_{max}}{M_0}$ { Point load	1.00,	1.08,	1.16,	1.27,	1.46,	1.79,	2.45,	4.47
UDL	1.00,	1.10,	1.20,	1.34,	1.58,	1.99,	2.83,	5.39

4 A pin-jointed strut length L carries an axial end load P. In addition there are couples (each M_0) applied at the ends, clockwise at one end and counterclockwise at the other. If $\alpha^2 = P/EI$ where EI is the flexural rigidity, show that the maximum bending moment in the strut is $M_0 \sec \alpha L/2$.

Obtain expressions for the maximum deflection (in terms of M_0, L, E and I) (a) when P is one-half of the Euler critical load and (b) when P is zero.
Answer (a) $0.254 M_0 L^2/EI$; (b) $M_0 L^2/8EI$ (compare with Example 6.1, Chapter 6 of Volume 1).

5 A tubular strut is 6 cm external and 5 cm internal diameter. It is 2 m long and has hinged ends. The load is parallel to the axis but eccentric to it. Find the maximum allowable eccentricity for a crippling load $0.75 \times$ Euler load, the yield stress being 300 MN/m^2. Work from first principles. $E = 206$ GN/m^2. *(UL)*
Answer 2.83 mm.

6 A long slender strut, originally straight and securely fixed at one end and (completely) free at the other end, is loaded at the free-end with an eccentric load whose line of action is parallel to the original axis of the strut. Deduce an expression for the deviation of the free end from its original position.

Determine this deviation and the greatest compressive stress for a steel strut complying with the above conditions; length 3 m; circular cross-section 5 cm external diameter and 2.5 cm internal diameter; load 3.6 kN and original eccentricity 7.5 cm. ($E = 206$ GN/m^2.) *(UL)*
Answer $e(\sec \alpha L - 1)$ where $\alpha^2 = P/EI$; 26.4 mm; 34.1 MN/m^2.

7 Derive a formula for the maximum compressive stress induced in an initially straight, slender, uniform strut when loaded along an axis having an eccentricity e at both ends, which are pin-jointed.

A straight steel pin-jointed strut is 5 cm diameter and 1.25 m long. Calculate:

(a) the Euler crippling load when loaded along the central axis;

(b) the eccentricity which will cause failure at 75 per cent of this load if the yield-point stress of the material is 270 MN/m^2. $E = 206$ GN/m^2. *(UL)*
Answer (a) 399 kN, (b) 1.01 mm.

8 A slender strut of uniform section and length L has ends which are fixed in position but free in direction. A force P acts at each end of the strut in a direction parallel with the axis of the strut and at a distance e from the axis but the two forces act on opposite sides of the neutral layer, i.e. their distance apart is $2e$.

Show that

(a) the condition for maximum bending moment is given by

$$\tan mx = -\cot \tfrac{1}{2}mL$$

where $m = \sqrt{(P/EI)}$ and x is the distance from an end to the section where maximum bending moment occurs, and

(b) maximum bending moment occurs at the ends if

$$P < \pi^2 EI/L^2 \qquad\qquad (UL)$$

Hint There are equal and opposite lateral reactions at the supports of $2Pe/L$. Since deflections are not required it is convenient to differentiate the bending moment equation twice and obtain d^2M/dx^2 as in Example 7.6.

9 A long strut of constant section is initially straight. A thrust is applied eccentrically at both ends and on the same side of the centre-line with the eccentricity at one end twice that at the other. The eccentricities are small compared with the length of the strut.

If the length of the strut is L and the thrust P, show that the maximum bending moment occurs at a distance x from the end with the smaller eccentricity where

$$\tan mx = \frac{2 - \cos mL}{\sin mL} \quad \text{and} \quad m = \sqrt{(P/EI)}$$

If in the above problem $L = 0.75$ m and the strut is circular and 2.5 cm diam., calculate the value of the eccentricities which will produce a maximum stress of 300 MN/m^2 with a thrust P of 35 kN. Take $E = 200$ GN/m^2.

Hint Differentiate bending moment equation twice as in Example 7.6.
Answer 2.8 mm and 5.6 mm.

10 (a) An initially straight, slender strut of uniform flexural rigidity EI and length l is built-in at both ends and loaded with an axial compressive load P. One end of the strut, and the force P acting at that end, are displaced laterally an amount e, so that they remain parallel to their original direction. If the end fixing moments, both of the same sense, are each denoted M, prove that

$$M = \frac{Pe(1 - \cos ml)}{2 - 2\cos ml - ml \sin ml}$$

where $m = \sqrt{(P/EI)}$.

(b) A steel tube 2.5 m long, having outside and inside diameters of 5 cm and $3\tfrac{1}{2}$ cm respectively, is used as a strut in the manner described in (a). If $P = 7000$ N and $e = 16$ mm find the value of M and also the maximum normal stress in the material. $E = 200\ \text{GN/m}^2$. *(UL)*

Answer 702 N m; 78.5 MN/m².

11 A hollow circular steel strut with its ends position-fixed has a length of 2.5 m, its external diameter being 10 cm and its internal diameter 8 cm. Before loading the strut is bent with a maximum deviation of 5 mm. Assuming that the centre line of the strut is sinusoidal, determine the maximum stress due to a central compressive end load of 100 kN. $E = 200\ \text{GN/m}^2$. *(UL)*

Answer 45.1 MN/m².

12 A slender strut 1.5 m long and of rectangular section 2.5 cm × 1 cm transmits a longitudinal thrust P acting at the centre at each end. The strut was slightly bent about its minor principal axis before loading.

If P is increased from 200 N to 600 N the deflection at the middle of the length increases by 1.4 mm; determine the amount of deflection before loading.

Find also the total deflection and the maximum stress when P is 1 200 N. $E = 205\ \text{GN/m}^2$. *(UL)*

Answer 4.0 mm; 11.1 mm and 36.8 MN/m² (compressive).

13 A straight circular section strut of length L has an applied axial compressive load P. It is loaded at the centre with a load W acting at right angles to its axis.

Prove that the maximum bending moment is

$$-\frac{W}{2m} \tan \tfrac{1}{2}mL \quad \text{where} \quad m = \sqrt{(P/EI)}$$

and derive a formula for the central deflection.

If the strut is of steel 2.5 cm dia. and 1.25 m long with an axial load of 16 kN calculate the value of W which will cause collapse if the yield point stress is 280 MN/m² and $E = 206\ \text{GN/m}^2$. *(UL)*

Answer $\dfrac{W}{2mP} \tan \tfrac{1}{2}mL - \dfrac{WL}{4P}$; 496 N

14 A straight horizontal rod of steel 3 m long and 3 cm in diameter is freely supported at its ends. An axial thrust of P is applied to each end. Find its greatest value if the maximum stress is not to exceed 36 MN/m². Density of steel 7.83 g/cm³; $E = 200\ \text{GN/m}^2$. *(UL)*

Hint A 'trial-and-error' solution is required; calculate P_E and take values of P between 0 and P_E.

Answer $P = 2\ 480\ \text{N}$.

15 A slender strut of uniform section and length L and with pin-jointed ends, is subjected to an axial thrust P and to a distributed lateral load of intensity increasing uniformly from zero at each end to a maximum of w per unit length at the mid-section of the strut. Show that the maximum bending moment is

$$M_{max} = \frac{w}{m^2}\left(\frac{2}{mL}\tan\tfrac{1}{2}mL - 1\right)$$

where $m^2 = P/EI$.

Further, show by using the approximation

$$\tan\theta = \theta + \theta^3/3 + 2\theta^5/15$$

that the value of M_{max} may be obtained by using the approximate formula

$$M_{max} = \mu + P\Delta$$

where μ is the maximum bending moment and Δ the maximum deflection due to the lateral load only. ($\Delta = wL^4/120EI$). (UL)

16 In the notation of the previous example show that the deflection at mid-span is

$$\frac{w}{P}\left(\frac{2}{m^3L}\tan\tfrac{1}{2}mL - \frac{L^2}{12} - \frac{1}{m^2}\right)$$

17 A pinned strut AB of constant section carries an axial load P and terminal couples M_A and M_B both deflecting the strut in the same direction. Obtain a formula for the maximum stress in the strut. (UL)

Answer

$$\frac{P}{A} + \frac{\sqrt{[M_A{}^2 + M_B{}^2 - 2M_A M_B \cos\alpha L]}}{Z\sin\alpha L}$$

where A and Z are the section area and modulus respectively and $\alpha^2 = P/EI$.

18 An initially straight slender strut of uniform section and length l has hinged ends through which it is axially loaded by a force P. The magnitude of P is between $\pi^2 EI/l^2$ and $4\pi^2 EI/l^2$. Equal opposite couples M_0 are applied to the ends to maintain equilibrium, the slope at each end then being ϕ radians. Show that

(a) $M_0 = \dfrac{P\phi}{m}(-\cot\tfrac{1}{2}ml)$

(b) the maximum deflection $\Delta = \dfrac{\phi}{m}\tan\tfrac{1}{4}ml$

where $m = \sqrt{(P/EI)}$.

If the strut is a steel bar 1 m long and of rectangular section 6 cm × 2 cm; $P = 2\pi^2 EI/l^2$ and $\phi = 0.01$ radian, find the values of M_0 and Δ. $E = 206$ GN/m². (UL)

Answer   (a) 280 N m; (b) 4.54 mm.

19 A strut of length $2a$ has each end fixed in an elastic material which exerts a restraining moment μ per unit angular displacement (one radian). Prove that the critical load P is given by the equation

$$\mu n \tan na + P = 0$$

where $n^2 = P/EI$.

Such a strut, 2.5 m in length, has a theoretical critical load of 15 500 N on the assumption that the ends are pinned, fixed in position but not in direction. Determine the percentage increase in the critical load if the constraint offered at the ends is 170 N m per degree of rotation. *(UL)*

Hint A 'trial-and-error' solution is required; take values of P between 1 and 4 times the critical load for pinned ends.

Answer 79.3 per cent.

20 An initially straight strut, L m long, has lateral loading w per metre and a longitudinal load P applied with eccentricity e m at both ends. If the strut has area A, relevant second moment of area I and section modulus Z, and the end moments and lateral loading have opposing effects, find an expression for the central bending moment, and show that the maximum stress at the centre will be equal to

$$\frac{P}{A} + \frac{1}{Z}\left[\left(Pe - \frac{wEI}{P}\right)\sec \tfrac{1}{2}L\sqrt{\frac{P}{EI}} + \frac{wEI}{P}\right] \qquad (UL)$$

Answer The central bending moment is given by the expression in square brackets.

21 A horizontal bar of uniform section and of length L, is simply supported at its ends. In addition to the uniform load w per unit length due to its own weight, the bar is subjected to longitudinal thrusts F acting at points on the vertical centre lines of the end sections at a distance e below the centres. Show that the resultant maximum bending moment in the beam will have its least possible value if

$$e = \frac{w(\sec \tfrac{1}{2}mL - 1)}{Fm^2(\sec \tfrac{1}{2}mL + 1)}$$

where $m^2 = F/EI$.

If the bar is of steel 2.5 m long of rectangular cross-section 8 cm wide and 2.5 cm deep and weighs 15.7 kgf/m and if the end thrust is 13 kN find the eccentricity e as already defined and also the corresponding maximum deflection. $E = 206$ GN/m^2. *(UL)*

Hint The required condition is obtained when the moment at the end is equal and opposite to that at mid-span. Equate $-Pe$ to the maximum bending moment as obtained in the previous example.

Answer 0.005 47 m or 5.47 mm; 0.59 mm (upwards).

22 A slender strut of length L is encastré at one end and pin-jointed at the other. It carries an axial load P and a couple M at the pinned end. If its flexural rigidity is EI and $P/EI = \mu^2$ show that the magnitude of the couple at the fixed end is

$$M\left(\frac{\mu L - \sin \mu L}{\mu L \cos \mu L - \sin \mu L}\right)$$

What is the value of this couple when P is one quarter of the Euler critical load and also when P is zero?   *(UL)*

Hint   The limiting value of $(\theta - \sin \theta)/(\theta \cos \theta - \sin \theta)$ as $\theta \to 0$ is found by expanding the trigonometrical functions as infinite series and considering the first few terms.

Answer $0.571M; \frac{1}{2}M$.

8

Beams of two materials

General considerations

This chapter deals with beams in which two materials (such as two dissimilar metals, steel and timber, or steel and concrete) are joined so that they deform together. The problems are analysed by modifying the simple bending theory introduced in Volume 1. In general, when two materials distort together their strains must be compatible.

Symmetrical cross-section

The two materials bend to the same radius of curvature and their separate moments of resistance can therefore be related.

Unsymmetrical cross-section

It is convenient to replace the composite section by an equivalent one of a single material having the same total moment of resistance for a given curvature.

Bimetallic strip

If a composite strip of two materials is heated the differential thermal expansion will cause internal stresses. At a cross-section of the strip the internal forces and moments of resistance must satisfy the conditions for equilibrium.

Reinforced concrete beam

Concrete is much weaker in tension than in compression and concrete beams are therefore reinforced by the addition of steel bars near the tension face (see Figure 8.4(a)). A simple analysis of such beams can be made on the basis of the following assumptions:

(1) all the tension is carried by the steel bars;
(2) the tensile stress in the steel is uniform;
(3) cross-sections of the beam remain plane during bending and there is no sliding of the steel bars relative to the concrete;
(4) Hooke's law holds for concrete in compression.

In fact, concrete departs from Hooke's law and to allow for this it is usual to take a lower value of the modulus of elasticity than that measured at low stresses.

Figure 8.4(b) shows the stress variation resulting from these assumptions. In general, the neutral axis does not pass through the centroid of the cross-section.

Worked examples

8.1 Moment of resistance and maximum stresses in flitched beam

> A flitched beam is made up of two timber joists each 100 mm wide by 240 mm deep, with a 20 mm steel plate 150 mm deep placed symmetrically between them and firmly attached to both. The plate is recessed into grooves cut in the inner faces of the joists so that the overall dimensions of the built-up section may be taken as 200 mm by 240 mm.
>
> Calculate the moment of resistance of the combined section when the maximum bending stress in the timber is 8.4 MN/m². What is then the maximum stress in the steel?
>
> Take E for steel as 206×10^9 N/m² and for timber as 12×10^9 N/m². (UL)

Solution   A flitched beam is a composite beam in which two timber joists and a steel plate are rigidly bolted together (Figure 8.1).

Since the steel and timber must bend to the same radius of curvature and, also, from the bending equation,

$$R = EI/M$$

then EI/M must have the same value for the steel and the timber. If suffixes s and t denote the steel and timber respectively,

$$\frac{E_s I_s}{M_s} = \frac{E_t I_t}{M_t} \tag{i}$$

It is convenient to calculate the I values in cm units and, with the dimensions given in the question,

$$I_s = \frac{2 \text{ cm} \times (15 \text{ cm})^3}{12} = 562.5 \text{ cm}^4$$

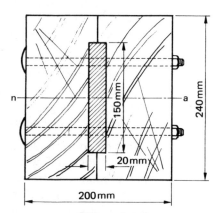

Figure 8.1

and

$$I_t = \frac{20\,\text{cm} \times (24\,\text{cm})^3}{12} - \frac{2\,\text{cm} \times (15\,\text{cm})^3}{12} = 22\,480\,\text{cm}^4$$

For the timber $y_{max} = 120\,\text{mm}$ and, for the given stress,

$$M_t = \frac{I_t}{y_{max}} \times \sigma$$

$$= \frac{(22\,480 \times 10^{-8}\,\text{m}^4) \times (8.4 \times 10^6\,\text{N/m}^2)}{120 \times 10^{-3}\,\text{m}}$$

$$= 15\,740\,\text{N m}$$

Thus from (i),

$$M_s = M_t \left(\frac{E_s I_s}{E_t I_t} \right)$$

$$= 15\,740\,\text{N m} \times \frac{(206 \times 10^9\,\text{N/m}^2) \times 562.5\,\text{cm}^4}{(12 \times 10^9\,\text{N/m}^2) \times 22\,480\,\text{cm}^4}$$

$$= 6\,760\,\text{N m}$$

Total moment of resistance is

$$M_t + M_s = (15\,740 + 6\,760)\,\text{N m} = 22.5\,\text{kN m} \qquad (Ans)$$

For the steel $y_{max} = 75\,\text{mm}$ and hence the maximum stress in the steel is

$$\sigma = \frac{M_s}{I_s} y_{max} = \frac{6\,760\,\text{N m}}{562.5 \times 10^{-8}\,\text{m}^4} \times 0.075\,\text{m}$$

$$= 90.1\,\text{MN/m}^2$$

8.2 Maximum stresses in timber beam of rectangular section with steel plate bolted to bottom surface

A timber beam 200 mm wide and 300 mm deep is reinforced by a steel plate 200 mm wide and 13 mm thick, bolted to its bottom edge, giving a composite beam 200 mm wide and 313 mm deep.

Explain how, for the purposes of calculation, the composite beam can be replaced by a steel beam of $\perp$ section. E for steel $= 20 \times E$ for timber.

Calculate the maximum stresses in the steel and timber when the composite beam carries a uniformly distributed load of 15 kN/m and is simply supported over a span of 6 m.

Solution The neutral axis of a composite beam does not in general pass through the centroid of the section. Since E for steel is much greater than E for timber then, for a given strain, the stress in the steel is much greater than that in the timber. Assuming that plane sections remain plane (for the composite beam) the longitudinal strain of a layer AB in the timber (Figure 8.2) is proportional to its distance y

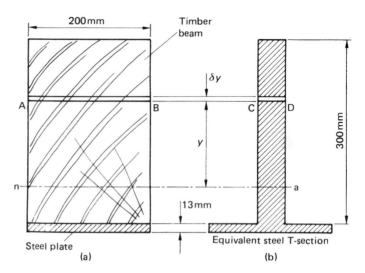

Figure 8.2

from the neutral axis. The corresponding stress acts on an area which, in the present example, is 200 mm wide. For the same radius of curvature, the stress in a steel strip CD, at the same distance y from the neutral axis, is much greater. In order that the steel strip CD shall carry the same longitudinal *force* as the timber strip AB, of the same thickness δy, it must be much narrower. In the present case, where E for steel $= 20 \times E$ for timber, the width of the equivalent steel section (where it replaces the timber) must be 1/20th of the width of the timber. The equivalent steel section has the same overall height and neutral axis position as the composite beam. For the same radius of curvature it has the same moment of resistance. The stress at any point in the steel is found in the usual way. The stress in the timber at any point is found by multiplying the stress at the corresponding point of the equivalent steel section by the ratio E timber/E steel.

In the example given, the timber beam, 200 mm wide, is replaced by a steel web whose width is

$$(E_t/E_s) \times 200 \text{ mm} = \tfrac{1}{20} \times 200 \text{ mm} = 10 \text{ mm}$$

the suffixes t and s referring to the timber and steel respectively.

The equivalent steel section is shown in Figure 8.2(b). By the usual methods (see Volume 1) it is found that its neutral axis is 87 mm from the bottom edge and the second moment of area about this axis is $I = 5\,638 \text{ cm}^4$.

The maximum bending moment on the beam is

$$M = \frac{WL}{8} = \frac{15 \text{ kN/m} \times 6 \text{ m} \times 6 \text{ m}}{8} = 67.5 \text{ kN m}$$

The maximum stress in the steel (of the composite beam) is equal to the stress at the bottom edge of the equivalent section and is

$$\sigma = \frac{My}{I} = \frac{67.5 \text{ kN m} \times (87 \times 10^{-3} \text{ m})}{5\,638 \times 10^{-8} \text{ m}^4}$$

$$= 104.2 \text{ MN/m}^2 \text{ (tensile)} \hspace{3cm} (Ans)$$

The stress at the top edge of the equivalent section is

$$\sigma = \frac{My}{I} = \frac{67.5 \text{ kN m} \times (226 \times 10^{-3} \text{ m})}{5\,638 \times 10^{-8} \text{ m}^4}$$

$$= 270.6 \text{ MN/m}^2 \text{ (compressive)}$$

The corresponding (maximum) stress in the timber is E_t/E_s times the amount, i.e.

$$\tfrac{1}{20} \times 270.6 \text{ MN/m}^2 = 13.5 \text{ MN/m}^2 \text{ (compressive)} \hspace{1.5cm} (Ans)$$

8.3 Deflection of bimetallic strip due to change in temperature

A bimetallic strip is composed of a brass strip 0.5 mm thick and 12 mm wide welded to a strip of steel 0.3 mm thick and also 12 mm wide. The strip is fixed at one end and has a length of 120 mm. If the strip is straight at 15°C calculate the end deflection when the temperature rises to 100°C. Work from first principles or prove any formula used.

Take the coefficients of expansion as 18×10^{-6} and 12.5×10^{-6} per degree C for brass and steel respectively and the values of E as 85 GN/m² and 205 GN/m² for brass and steel respectively

(UL)

Solution If the brass strip is uppermost, its greater coefficient of expansion will lead to circular arc bending as shown in Figure 8.3(a) when the temperature rises. The radius of the deflection curve R is so large compared with the cross-sectional dimensions of the strip that it may be taken as the same for both materials.

Suppose the coefficients of expansion for brass and steel are α_b and α_s respectively, E_b and E_s being the corresponding values of the moduli of elasticity.

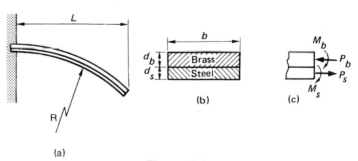

(a)

(b)

(c)

Figure 8.3

Figure 8.3(b) shows an enlarged view of the cross-section and, as shown in 8.3(c), the internal force system for each material can be reduced to a central force and a bending moment.

Since the net longitudinal force is zero we can put

$$P_b = P_s = P$$

The longitudinal forces in the separate materials are therefore equal and opposite, the distance between them being $\frac{1}{2}(d_b + d_s)$.

The moment of this couple must, for equilibrium, balance the sum of the bending moments of resistance. Thus

$$\frac{P}{2}(d_b + d_s) = M_b + M_s$$

$$= \frac{E_b I_b}{R} + \frac{E_s I_s}{R}$$

or

$$\frac{PR}{2} = \frac{E_b I_b + E_s I_s}{d_b + d_s} \tag{i}$$

A second relationship can be obtained by equating the longitudinal strains of the two materials at their common surface. These strains are made up of three components, the temperature expansion, the strain due to the bending moment M and the strain due to the longitudinal force P.

From the bending equation $\sigma/y = E/R$ and hence the bending strain may be written as

$$\frac{\sigma}{E} = \frac{y}{R} = \frac{d}{2R}$$

Thus, net strain for brass = net strain for steel and, for a temperature rise t,

$$\alpha_b t - \frac{P}{bd_b} \times \frac{1}{E_b} - \frac{d_b}{2R} = \alpha_s t + \frac{P}{bd_s} \times \frac{1}{E_s} + \frac{d_s}{2R}$$

or

$$\frac{P}{b}\left(\frac{1}{d_b E_b} + \frac{1}{d_s E_s}\right) + \frac{1}{2R}(d_b + d_s) = (\alpha_b - \alpha_s)t \tag{ii}$$

The simultaneous equations (i) and (ii) may be solved to determine the values of the required quantities. In the present case the unknowns are P and R and substituting for the other quantities we have (working in metres),

$$\frac{PR}{2} = \frac{85 \times 10^9 \times \left(\dfrac{0.012 \times 0.000\,5^3}{12}\right) + 205 \times 10^9 \times \left(\dfrac{0.012 \times 0.000\,3^3}{12}\right)}{0.000\,5 + 0.000\,3}$$

or

$$PR = 20.2 \tag{iii}$$

and

$$\frac{P}{0.012} \left(\frac{1}{0.000\,5 \times 85 \times 10^9} + \frac{1}{0.000\,3 \times 205 \times 10^9} \right)$$

$$+ \frac{1}{2R} (0.000\,5 + 0.000\,3) = (18 \times 10^{-6} - 12.5 \times 10^{-6}) \times 85$$

and

$$3.316 \times 10^{-6}P + \frac{400 \times 10^{-6}}{R} = 467.5 \times 10^{-6}$$

or

$$3.316P + \frac{400}{R} = 467.5 \tag{iv}$$

Substituting in (iv) for P from (iii)

$$3.316 \times \frac{20.2}{R} + \frac{400}{R} = 467.5$$

from which $R = 0.999$ m

For circular arc bending (see Volume 1, Chapter 6), the deflection is given by

$$\delta = \frac{L^2}{2R} = \frac{0.120^2}{2 \times 0.999}$$

$$= 0.007\,21 \text{ m or } 7.21 \text{ mm} \qquad (Ans)$$

8.4 Rectangular concrete beam with tension reinforcement

A rectangular reinforced-concrete beam is reinforced in tension only; its effective depth is 500 mm and its width is 250 mm. Permissible stresses are 7 and 140 MN/m² for concrete and steel respectively, and the modular ratio is 15.

Calculate the moment of resistance of the beam if the area of tensile steel provided is (a) 1 000 mm² and (b) 2 500 mm². *(UL)*

Solution The analysis is based on the assumptions listed at the beginning of the chapter and the notation is illustrated in Figure 8.4. Let

$n =$ distance of neutral axis from top (compression) face
$b =$ width of the beam
$d =$ depth to the centre of the steel bars (the 'effective' depth)
$A_s =$ total cross-sectional area of the steel
$\sigma_s =$ uniform tensile stress in the steel
$\sigma_c =$ maximum compressive stress in the concrete
$E_s, E_c =$ moduli of elasticity for steel and concrete respectively.

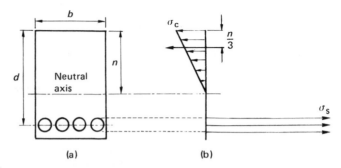

Figure 8.4

With this notation the resultant compressive force in the concrete is given by taking the average stress $\frac{1}{2}\sigma_c$ over the area above the neutral axis, $b \times n$. Equating this force to the tensile force in the steel, we have

$$\frac{1}{2}\sigma_c bn = \sigma_s A_s \qquad \text{(i)}$$

The maximum strain in the concrete is σ_c/E_c and the strain in the steel is σ_s/E_s. Since cross-sections are assumed to remain plane in bending these strains are proportional to the corresponding distances from the neutral axis. Thus

$$\frac{\sigma_c}{E_c n} = \frac{\sigma_s}{E_s(d-n)}$$

or

$$\frac{m\sigma_c}{\sigma_s} = \frac{n}{d-n} \qquad \text{(ii)}$$

where m = the modular ratio E_s/E_c.

Note that the stresses given in the question are permissible values and will not, in general, be reached simultaneously. The way in which equations (i) and (ii) are used depends upon the data given and the answers required.

In the present case, we need to locate the neutral axis and, substituting for the ratio σ_c/σ_s in (ii) its value given in (i)

$$\frac{m \times 2A_s}{bn} = \frac{n}{d-n}$$

from which the following quadratic for n is obtained

$$bn^2 + 2mA_s n - 2mA_s d = 0$$

(a) Working in metres, $b = 0.25$, $d = 0.5$, $A_s = 0.001$ and, with $m = 15$, the equation becomes

$$0.25n^2 + 0.03n - 0.015 = 0$$

and the solution is $n = -0.312$ or 0.192. Taking the positive root, the neutral axis is 0.192 m (or 192 mm) from the compression face.

The actual stresses in the steel and concrete (as distinct from the permissible values) are related by equation (i) or (ii). Using (i) and starting with the maximum permissible stress in the steel we have

$$\sigma_c = \frac{2\sigma_s A_s}{bn} = \frac{2 \times (140 \times 10^6) \times (1\,000 \times 10^{-6})}{0.250 \times 0.192}$$

$$= 5.83 \times 10^6 \, \text{N/m}^2 \text{ or } 5.83 \, \text{MN/m}^2$$

This is less than the permitted value and hence the steel is the limiting factor. Hence, tensile force in steel

$$= (140 \times 10^6) \times (1\,000 \times 10^{-6})$$

$$= 140 \, \text{kN}$$

and this is also the resultant compressive force in the concrete. Since the stress distribution diagram for the concrete is triangular the latter force acts at $n/3$ from the top face as shown in Figure 8.4(b). The two forces form a couple and the distance between their lines of action is

$$\left(d - \frac{n}{3}\right)$$

Hence, moment of resistance

$$= \text{force in steel} \times \left(d - \frac{n}{3}\right)$$

$$= (140 \times 10^3) \times \left(0.5 - \frac{0.192}{3}\right)$$

$$= 61.0 \, \text{kN m} \hspace{4cm} (Ans)$$

(b) With $A_s = 0.002\,5 \, \text{m}^2$ the quadratic equation becomes

$$0.25n^2 + 0.075n - 0.037\,5 = 0$$

and the positive root is $n = 0.265$. The neutral axis is therefore 265 mm from the compression face.

Using the maximum permissible stress in the steel, as before, the corresponding maximum stress in the concrete is

$$\sigma_c = \frac{2\sigma_s A_s}{bn} = \frac{2 \times (140 \times 10^6) \times (2\,500 \times 10^{-6})}{0.250 \times 0.265}$$

$$= 10.57 \, \text{MN/m}^2$$

This exceeds the permitted value of $7 \, \text{MN/m}^2$ and hence the concrete is the limiting factor. The actual stress in the steel is given by

$$\sigma_s = \frac{\sigma_c bn}{2A_s} = \frac{(7 \times 10^6) \times 0.250 \times 0.265}{2 \times (2\,500 \times 10^{-6})}$$

$$= 92.75 \, \text{MN/m}^2$$

With this value the tensile force in the steel is

$$(92.75 \times 10^6) \times (2\,500 \times 10^{-6}) = 231.9 \text{ kN}$$

$$\text{and moment of resistance} = (231.9 \times 10^3) \times \left(0.5 - \frac{0.265}{3}\right)$$

$$= 95.5 \text{ kN m} \qquad \qquad (Ans)$$

8.5 Area of steel required for balanced reinforcement of singly reinforced concrete beam

> A singly reinforced concrete beam is 300 mm wide and 225 mm deep with cover such that the steel is placed 175 mm from the compression edge and 50 mm from the tensile edge. Find the cross-sectional area of steel required to give 'balanced reinforcement' and the corresponding moment of resistance if the allowable stresses in the concrete and steel are 4 MN/m² compression and 100 MN/m² tension respectively. The ratio of Young's modulus for steel to that of concrete is 15. (UL)

Solution The cross-section of the beam is shown in Figure 8.5.

(Single reinforcement implies reinforcement for tension only; in practice, steel bars are often placed near the compression face also.) As in the previous example it is assumed that the concrete contributes nothing in tension and the overall depth of the beam does not enter the calculations. However, the 'cover' of concrete at the bottom of the section is needed to ensure composite action of the two materials and to provide fire protection for the steel.

In the present example the area of steel can be chosen to ensure that the permissible stresses in the two materials are achieved simultaneously. In this condition the beam is said to have *balanced reinforcement*.

With the figures given in the question $d = 0.175$, $m = 15$ and the ratio of permitted stresses is

$$\frac{\sigma_c}{\sigma_s} = \frac{4 \times 10^6}{100 \times 10^6} = \frac{1}{25}$$

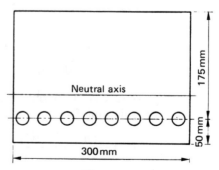

Figure 8.5

Hence from equation (ii) of the previous example

$$\frac{15}{25} = \frac{n}{0.175 - n}$$

from which

$$n = 0.656 \text{ m} \text{ or } 65.6 \text{ mm}$$

Using this result and equation (i) of the previous example we have

$$A_s = \frac{\sigma_c bn}{2\sigma_s}$$

$$= \frac{0.300 \times 0.065\,6}{2 \times 20}$$

$$= 492 \times 10^{-6} \text{ m}^2 \text{ or } 492 \text{ mm}^2 \qquad (Ans)$$

With this result,

$$\text{force in steel} = \text{stress} \times \text{area}$$

$$= (100 \times 10^6) \times (492 \times 10^{-6})$$

$$= 49.2 \text{ kN}$$

and moment of resistance $= \text{force} \times \left(d - \frac{n}{3} \right)$

$$= 49.2 \times 10^3 \times \left(0.175 - \frac{0.065\,6}{3} \right)$$

$$= 7.53 \text{ kN m} \qquad (Ans)$$

8.6 Safe bending moment for reinforced concrete T-beam

> A reinforced concrete T-beam has a flange 900 mm wide by 75 mm deep, the stem width being 300 mm. The tensile reinforcement is 2 000 mm² in area and 600 mm below the upper surface of the flange. The safe stresses are 120 MN/m² and 7 MN/m² in steel and concrete respectively and the modular ratio is 15. Find the bending moment that this section will safely resist.
>
> Mention the factors which limit the effective width of flange of such a beam and comment on the reasons. (UL)

Solution The cross-section is shown in Figure 8.6(a) and the stress distribution in Figure 8.6(b). At the outset, the position of the neutral axis is unknown; if it were within the flange, the analysis would be identical to that for a rectangular beam of 900 mm width since it is assumed that the concrete makes no contribution in tension. This means that (theoretically) the width of the section can be reduced below the neutral axis without detracting from the strength.

Assuming, however, that the neutral axis lies within the 'stem' or web as shown, then the stress in the concrete at the intersection of the

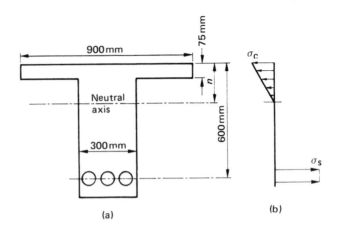

Figure 8.6

stem and flange is, by proportion

$$\left(\frac{n-0.075}{n}\right)\sigma_c \quad \text{or} \quad \left(1-\frac{0.075}{n}\right)\sigma_c$$

and the average compressive stress in the concrete below this line is one-half this amount.

The average compressive stress in the flange is

$$\frac{1}{2}\left[\left(1-\frac{0.075}{n}\right)\sigma_c+\sigma_c\right] \quad \text{or} \quad \left(1-\frac{0.037\,5}{n}\right)\sigma_c$$

Working in metre units and equating the total compressive force in the concrete to the tensile force in the steel, we have

$$\frac{1}{2}\left(1-\frac{0.075}{n}\right)\sigma_c\times0.300\times(n-0.075)$$

$$+\left(1-\frac{0.037\,5}{n}\right)\sigma_c\times0.900\times0.075=2\,000\times10^{-6}\sigma_s$$

which reduces to

$$\frac{\sigma_s}{\sigma_c}=75n+22.5-\frac{0.843\,5}{n} \tag{i}$$

The strain in the steel is σ_s/E_s. These strains are proportional to the corresponding distances from the neutral axis, giving

$$\frac{\sigma_s E_c}{E_s \sigma_c}=\frac{0.600-n}{n}$$

and with the modular ratio of 15 this leads to

$$\frac{\sigma_s}{\sigma_c}=\frac{15(0.600-n)}{n} \tag{ii}$$

Eliminating the ratio σ_s/σ_c from (i) and (ii) gives

$$\frac{15(0.600 - n)}{n} = 75n + 22.5 - \frac{0.843\,5}{n}$$

which, on rearrangement, leads to the quadratic equation

$$n^2 + 0.5n - 0.131\,3 = 0$$

the positive root being $n = 0.190$ (m). The neutral axis is therefore 190 mm from the compression face and is below the flange/stem intersection.

From (ii) the ratio of stresses is

$$\frac{\sigma_s}{\sigma_c} = \frac{15(0.600 - 0.190)}{0.190} = 32.4$$

This is greater than the ratio of the permissible stresses and hence the limiting stress in the steel, 120 MN/m^2, is the critical factor. The corresponding stress in the concrete is

$$\sigma_c = \frac{\sigma_s}{32.4} = \frac{120 \times 10^6}{32.4} = 3.70 \text{ MN/m}^2$$

By proportion, the corresponding stress in the concrete at the flange/stem junction is

$$3.70 \times 10^6 \times \left(\frac{0.190 - 0.075}{0.190}\right) = 2.24 \text{ MN/m}^2$$

The stress variation through the flange can be considered in two parts, a uniform stress of 2.24 MN/m^2 and a stress which varies linearly from zero at the junction to $(3.70 - 2.24)$, i.e. 1.46 MN/m^2 at the top face.

The compressive force can now be subdivided into three parts as follows, each calculated as the product of an average stress and the corresponding area. The results are

force in stem $= \frac{1}{2} \times (2.24 \times 10^6) \times 0.300 \times (0.190 - 0.075)$

$= 38.6 \text{ kN}$

force in flange due to uniform stress of 2.24 MN/m^2

$= 2.24 \times 10^6 \times 0.900 \times 0.075$

$= 151.2 \text{ kN}$

force in flange due to 'additional' stress which varies from zero at junction to 1.46 MN/m^2 at top face and this equals

$\frac{1}{2} \times 1.46 \times 10^6 \times 0.900 \times 0.075$

$= 49.3 \text{ kN}$

The distances of the lines of action of these forces from the neutral axis are respectively

$$\tfrac{2}{3} \times (0.190 - 0.075) = 0.076\,7 \text{ m}$$

$$0.190 - \frac{0.075}{2} = 0.152\,5 \text{ m}$$

and

$$0.190 - \frac{0.075}{3} = 0.165\,0 \text{ m}$$

Hence the resultant moment about the neutral axis of the compressive forces in the concrete is

$$38.6 \times 0.076\,7 + 151.2 \times 0.052\,5 + 49.3 \times 0.165\,0 = 34.2 \text{ kN m}$$

The tensile force in the steel is given by

$$\text{stress} \times \text{area} = (120 \times 10^6) \times (2\,000 \times 10^{-6})$$
$$= 240 \text{ kN}$$

a result that tallies with the total compressive force in the concrete. Its line of action is $(0.600 - 0.190)$, i.e. 0.410 m from the neutral axis and its moment about this axis is, therefore,

$$240 \times 0.410 = 98.4 \text{ kN m}$$

Adding the results for the concrete and steel gives

$$\text{maximum safe bending moment} = 34.2 + 98.4$$
$$= 132.6 \text{ kN m} \qquad (Ans)$$

It is interesting to compare this result with that for a rectangular beam having the same permissible stresses, effective depth and area of reinforcement but with a width of 900 mm throughout. It can be shown by the methods used in solving Example 8.4 that the neutral axis of such a beam is 169 mm from the top (compressive) face and its maximum safe bending moment is 130.5 kN m. The T-section therefore achieves a slightly higher safe bending moment with less than half as much concrete.

In the design of reinforced concrete beams, T-sections are often preferred to rectangular. However, the flange width cannot be increased indefinitely because the outermost parts of side flanges do not satisfy the assumptions used in the present analysis and there is a lateral variation in stress. Furthermore, if the load on the beam is distributed over the top face of the flange there may be bending action in a plane at right angles to that of the longitudinal bending.

For these reasons flange width is restricted in design codes. For example, the British Standard code of practice CP110 'The Structural Use of Concrete' (Part I, 1972) states that the effective width for a T-beam should not exceed the lesser of

– the width of the web plus one-fifth of the distance between the points of zero bending moment, or
– actual width of the flange.

Problems

1 A composite beam consists of a timber joist 240 mm deep × 150 mm wide with a steel plate 150 mm × 10 mm bolted on each side, the steel plates being symmetrical about the axis of bending. If the stresses in the timber and steel are not to exceed 7 MN/m² and 140 MN/m² respectively, find the maximum bending moment the beam will carry and the maximum stresses in the two materials when carrying this moment.
Compare the value of this moment with that for the timber joist alone.

$$E(\text{steel}) = 206 \text{ GN/m}^2 \qquad E(\text{timber}) = 10 \text{ GN/m}^2$$

(IStructE)

Answer Maximum bending moment is 16.84 kN m, the stresses in the timber and steel being 7 MN/m² and 90 MN/m² respectively. This moment is 1.67 times that for the timber joist alone.

2 Figure 8.7 shows the cross-section of a compound bar which has been formed by brazing a steel strip 40 mm by 10 mm to a brass strip 40 mm by 20 mm. The compound bar is bent to a circular arc with neutral axis parallel to AB. If E for brass is 82 GN/m² and E for steel is 206 GN/m², find (a) the position of the neutral axis, and (b) the ratio of the maximum stress in the steel to that in the brass. Also find the radius of curvature of the neutral surface when the maximum stress in the steel becomes 70 MN/m². *(IMechE)*
Answer Neutral axis is 11.65 mm from AB. The ratio of the maximum stress in the steel to that in the brass is 1.60. Required radius of curvature is 34.3 m.

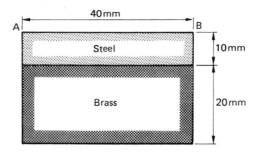

Figure 8.7

3 A timber beam 80 mm wide by 160 mm deep is to be reinforced by bonding strips of aluminium alloy 80 mm wide on to the top and bottom faces of the timber over the whole length of the beam. If the moment of resistance of the composite beam is to be four times that of the timber alone and the same value of the maximum bending stress in the timber is reached in both cases, determine the thickness of the alloy strip and the ratio of the maximum bending stresses in the alloy strip and timber. E for alloy strip $= 7.15 \times E$ for timber. (UL)

Answer Required thickness is 9.92 mm (top and bottom). The maximum bending stress in the alloy is 8.035 times that in the timber.

4 A timber beam 100 mm wide by 200 mm deep is to be reinforced with two steel plates 16 mm thick. Compare the moments of resistance for the same value of the maximum bending stress in the timber when the plates are: (i) 100 mm wide and fixed to the top and bottom surfaces of the beam; or (ii) 200 mm deep and fixed to the vertical sides of the beam. (E for steel $= 20 \times E$ for timber.) (UL)

Answer The moment of resistance in case (i) is 1.416 times that in case (ii).

5 Two rectangular bars, one of brass and the other steel, each 36 mm by 10 mm are placed together to form a beam 36 mm wide and 20 mm deep, on two supports 0.8 m apart. The brass bar is on top of the steel.

Determine the maximum central load which can be applied to the beam, if the bars are (a) separate and can bend independently, (b) firmly secured to each other throughout their length.

For brass $E = 86 \times 10^9$ N/m², for steel $E = 206 \times 10^9$ N/m².

Maximum allowable stress in brass is 70 MN/m² and in steel 105 MN/m². (UL)

Hint In (a) assume that the bars bend to the same radius of curvature and use the method of Example 8.1; in (b) find the equivalent steel beam as in Example 8.2.

Answer (a) 446 N; (b) 980 N; steel is limiting factor in both cases.

6 A brass strip 50 mm by 13 mm in section is riveted to a steel strip 64 mm $\times$ 10 mm in section to form a compound beam of total depth 23 mm, the brass strip being on top and the beam section being symmetrical about the vertical axis. The beam is simply supported on a span of 1.2 m and carries a load of 1.6 kN at mid-span.

(a) Determine the maximum stresses in each of the materials due to bending;

(b) make a diagram showing the distribution of bending stress over the depth of the beam;

(c) determine the maximum deflection.

Take E for steel $= 206$ GN/m²; E for brass $= 103$ GN/m². (UL)

Answer (a) In steel 110.9 MN/m² (tensile); in brass, 90.1 MN/m² (compressive); (c) 7.3 mm.

7 A straight bimetallic strip consists of a strip of brass of rectangular section of width b and thickness t joined along the whole of its length to a strip of steel of the same dimensions, thus forming a composite bar of width b and thickness $2t$. If the bar is uniformly heated and is quite free to bend, show that it will bend to a radius

$$R = \frac{E_B^2 + E_S^2 + 14E_B E_S}{E_B E_S(\alpha_B - \alpha_S)} \cdot \frac{t}{T}$$

where E_B and E_S are the moduli of elasticity of brass and steel respectively, α_B and α_S the corresponding coefficients of linear expansion, T = rise in temperature.

A bimetallic strip 200 mm long with the steel and brass strips each 1.5 mm thick rests on a truly level surface with the brass uppermost. If the strip is initially straight, find the maximum clearance between it and the surface due to a temperature rise of 60°C. Take

$$\alpha_B = 19 \times 10^{-6} \text{ per °C} \quad \text{and} \quad \alpha_S = 11 \times 10^{-6} \text{ per °C}$$

$$E_B = 97 \text{ GN/m}^2 \qquad \text{and} \quad E_S = 205 \text{ GN/m}^2 \qquad (UL)$$

Answer 0.096 mm.

8 For the bimetallic strip of the previous example determine the maximum stresses in the brass and the steel.
Hint The result is obtained by adding the stress due to the longitudinal force to the maximum stress due to the curvature.
Answer Brass, 25.6 MN/m²; steel, 44.3 MN/m².

9 Design a singly-reinforced concrete beam to carry 20 kN/m over a span of 8 metres. The depth of the section to the steel reinforcement is D, the breadth $D/2$, the permissible stresses in concrete and steel are 7 MN/m² and 212 MN/m² respectively and the modular ratio is 13.

Subsequently it is found necessary to limit the permitted steel stress in the beam which has been constructed according to the previous calculation. The permitted stress is now 140 MN/m², what is the permitted loading?

Describe briefly the principal features of the limit state method of design adopted in the current draft code of practice for the structural use of concrete. (UL)
Note The last part of the question is not covered in this book. Reference should be made to the specialized texts on reinforced concrete design.
Answer On the basis of balanced reinforcement $D = 697$ mm and the required area of steel is 1 204 mm². With the reduced stress in the steel the permissible loading is 13.2 kN/m.

10 A singly-reinforced concrete beam of rectangular section, width b, is reinforced with steel bars of total cross-sectional area

A at a depth *d* below the compression face. If the effective steel/concrete area ratio *A/bd* is denoted by *r*, show that the distance *n* of the neutral axis below the compression face is given by

$$\frac{n}{d} = \sqrt{[(rm)^2 + 2rm - rm]}$$

where *m* is the modular ratio.

If, for such a beam, the stress in the steel is not to exceed 90 MN/m², the width *b* and effective depth *d* are 800 mm and 400 mm respectively, and the area of the steel is 3 per cent of the area of the concrete, what is the bending moment that the beam can withstand and what is the corresponding greatest stress in the steel? Take the modular ratio as 15.
Answer   553 kN m; 9 MN/m².

11   A singly-reinforced rectangular concrete beam, width *b*, effective depth *d* and area of steel *A* is designed for maximum permissible stresses in the steel and concrete of σ_s and σ_c respectively. Show that the *economic ratio*, i.e. the value of *A/bd* for balanced reinforcement, is given by

$$\frac{1}{\dfrac{2}{m}\left(\dfrac{\sigma_s}{\sigma_c}\right)^2 + \dfrac{\sigma_s}{\sigma_c}}$$

and calculate its value when $\sigma_s = 20\sigma_c$ and $m = 15$.

Determine the dimensions of such a beam, and the corresponding area of steel, to withstand a bending moment of 500 kN m if the permitted stress in the steel is 120 MN/m². Take *d* = 2*b*.
Answer   0.013 6; 447 mm wide and 894 mm deep.

12   A rectangular singly-reinforced beam is to have an effective depth of twice the breadth and is required to carry a uniformly distributed load of 20 kN per metre run (including weight of beam) over a simply-supported span of 8 m together with a single concentrated load of 50 kN at a distance of 2.4 m from one support. If the allowable compressive stress in the concrete is 7 MN/m², the allowable tensile stress in the steel is 120 MN/m² and the modular ratio is 15, determine the required dimensions of the beam and the area of steel reinforcement. Work from first principles or prove any formula used in the design of the beam section stating the assumptions.   (*UL*)
Answer   With balanced reinforcement, effective depth = 690 mm, width = 345 mm and area of reinforcement = 3 227 mm².

13   A rectangular reinforced concrete beam is to be designed to carry an inclusive uniform load of 9 kN per metre run on an effective span of 6 m the maximum bending moment being

WL/10. The breadth of the beam is two-thirds of the effective depth and the steel reinforcement is to be 'economic'. Calculate from first principles, stating the assumptions, the size of the beam and the area of steel reinforcement assuming a modular ratio of 12 and allowable stresses of 5 MN/m^2 and 80 MN/m^2 for the concrete and steel respectively.   (*UL*)
Answer Effective depth = 375 mm, width = 250 mm, reinforcement area = 1260 mm^2.

14 A reinforced concrete T-beam has an effective depth of 400 mm. The tension steel at the centre of the span consists of six 18 mm diameter bars. The rib breadth is 200 mm, the breadth of the compression flange is 900 mm and its thickness is 100 mm. Calculate the moment of resistance of the section and the actual maximum stresses corresponding to this moment.
   Take the permissible stresses as 5 MN/m^2 compression and 120 MN/m^2 tension. *m* = 18.   (*UL*)
Answer 65.8 kN m; 3.25 MN/m^2 and 120 MN/m^2.

15 The flange of a reinforced concrete T-beam is 750 mm wide and the rib is 250 mm wide. Calculate the moment of resistance of the beam if it has 1 500 mm^2 of steel centred 400 mm below the top of the flange,

   (a) when the thickness of the flange is 75 mm
   (b) when the thickness of the flange is 150 mm.

The permissible stresses in steel and concrete are 120 MN/m^2 and 5 MN/m^2 respectively and *m* = 18. Prove any special formula used.   (*UL*)
Answer Stress in steel is limiting factor.

   (a) 64.9 kN m   (b) 63.75 kN m.

9

Special bending and torsion problems

Plastic bending

It is assumed that the material remains elastic up to the yield point and that further straining takes place at a constant stress. If the cross-section is unsymmetrical the neutral axis does not coincide with that in simple bending. When the section of a beam reaches a fully-plastic state, a 'plastic hinge' is formed. The ratio of the moment of resistance under this condition to that at which yielding commences is called the 'shape factor'.

Unsymmetrical bending

The bending stress at a given point in the section is the resultant of the stresses due to bending about the two principal axes. The determination of the position of the principal axes and the values of I (the second moment of area) about them is discussed in Appendix 1.

The bending stress at a point whose co-ordinates are (x, y) with respect to central axes XX and YY which are mutually perpendicular but not principal axes is

$$\sigma = \frac{(M_{XX}I_{YY} - M_{YY}I_{XY})y + (M_{YY}I_{XX} - M_{XX}I_{XY})x}{I_{XX}I_{YY} - I_{XY}{}^2}$$

Bending of flat plates

For a symmetrically loaded circular plate the differential equation of flexure is

$$\frac{d}{dr}\left[\frac{1}{r}\frac{d(r\theta)}{dr}\right] = -\frac{F}{D}$$

where F = shearing force per unit circumferential length at radius r,
$\theta = dy/dr$, y being the deflection at radius r,
$D = Et^3/12(1 - v^2)$, the flexural rigidity, where E = Young's modulus of elasticity, t = plate thickness and v = Poisson's ratio.

If F is expressed in terms of r, the equation can be integrated twice to find θ, the slope, and again to find y, the deflection.

The bending moment M in a radial plane per unit circumferential length is given by

$$M = D\left(\frac{d\theta}{dr} + \frac{v\theta}{r}\right)$$

Heavily curved beams   The neutral axis does not pass through the centroid of the section. If e is the distance of the neutral axis from the centroid (positive when the neutral axis is between the centroid and the centre of curvature)

$$e = \frac{1}{A} \int \frac{y^2}{R_0 + y} \, dA$$

where A = area of the cross-section

y = distance from the neutral axis, positive outwards

R_0 = initial radius of the neutral surface.

The bending stress σ is given by

$$\sigma = \frac{My}{Ae(R_0 + y)}$$

where M is the bending moment on the section.

Worked examples

9.1 Plastic bending of simply-supported rectangular beam with central point load

A rectangular steel beam, width b and depth d, is simply supported over a span L and a load W is gradually applied at mid-span. The steel follows a linear stress–strain law up to a yield stress σ_Y; at this constant stress considerable plastic deformation occurs. It may be assumed that the properties of the steel are the same in tension and compression.

Calculate the value of W at which

(a) yielding commences,
(b) yielding penetrates half-way to the neutral axis,
(c) a plastic hinge is formed.

What is the value of the shape factor for a rectangular section?

Solution   The stress–strain relationship described in the question is illustrated by Figure 9.1. It ignores any drop in stress at yield and

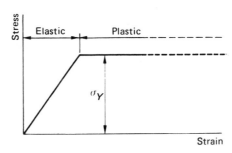

Figure 9.1

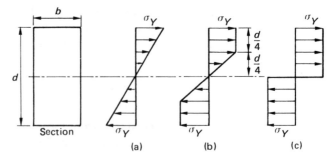

Figure 9.2

assumes that the material undergoes considerable strain at constant stress.

In applying this relationship to the bending of beams it is further assumed that plane cross-sections remain plane and that the same conditions apply in compression as in tension.

(a) Up to the load at which yielding commences the simple theory of bending applies and the distribution of stress over the section is shown in Figure 9.2(a). For the cross-section $I = bd^3/12$, the distance from the neutral axis to the edge of the section is $\frac{1}{2}d$ and, from the bending equation, the moment of resistance when yielding commences is

$$M = \frac{\sigma I}{y} = \frac{\sigma_Y \times (bd^3/12)}{\frac{1}{2}d} = \frac{1}{6}\sigma_Y bd^2$$

The maximum bending moment for a centrally-loaded beam simply supported at its ends is $\frac{1}{4}WL$ and hence

$$\frac{1}{4}WL = \frac{1}{6}\sigma_Y bd^2 \quad \text{from which} \quad W = 2\sigma_Y bd^2/3L \qquad (Ans)$$

(b) If the load is increased yielding penetrates from the extreme fibres towards the neutral axis. Although the outer fibres are strained more and more the maximum stress remains at the value σ_Y and the stress distribution is of the form shown in Figure 9.2(b). The central portion of the cross-section remains elastic and its moment of resistance can be calculated using simple bending theory and a reduced depth. If yielding penetrates half-way to the neutral axis, $y = \frac{1}{2}d$. The corresponding bending moment is

$$M(\text{elastic}) = \frac{\sigma_Y \frac{1}{12}b(\frac{1}{2}d)^3}{\frac{1}{4}d} = \frac{\sigma_Y bd^2}{24}$$

For the plastic portion the stress σ_Y is constant and there are forces, each $\sigma_Y \times b \times \frac{1}{4}d$, acting at distances $\frac{3}{8}d$ from the neutral axis. These produce a moment

$$M(\text{plastic}) = 2\sigma_Y \times \frac{1}{4}bd \times \frac{3}{8}d = \frac{3\sigma_Y bd^2}{16}$$

Thus, total moment of resistance is

$$M = M(\text{elastic}) + M(\text{plastic})$$
$$= \sigma_Y bd^2/24 + 3\sigma_Y bd^2/16$$
$$= 11\sigma_Y bd^2/48$$

The corresponding load is

$$W = 11\sigma_Y bd^2/12L \qquad\qquad (Ans)$$

(c) If the load is further increased plastic deformation continues until yielding penetrates to the neutral axis as indicated in Figure 9.2(c). (A very small elastic region remains near the neutral axis but its contributions to the moment of resistance may be neglected.)

The corresponding forces above and below the neutral axis are each $\sigma_Y \times b \times \frac{1}{2}d$ and act at distances $\frac{1}{4}d$ from it. Thus the moment of resistance is

$$M = 2 \times \sigma_Y \times \tfrac{1}{2}bd \times \tfrac{1}{4}d = \tfrac{1}{4}\sigma_Y bd^2$$

and the corresponding load is

$$W = \sigma_Y bd^2/L \qquad\qquad (Ans)$$

At this load a plastic 'hinge' is formed at mid-span and any further increase would cause collapse through the rotation of the halves of the beam about its mid-point.

Shape factor is a property of the cross-section and is defined as the ratio of the moment of resistance in the fully plastic state to that at which yielding commences.

Thus, using the present results,

$$\text{shape factor} = \frac{\text{moment in case (c)}}{\text{moment in case (a)}}$$
$$= \frac{\sigma_Y bd^2}{4} \bigg/ \frac{\sigma_Y bd^2}{6} = \frac{3}{2} \qquad\qquad (Ans)$$

9.2 Plastic bending and collapse of fixed-ended beam with uniformly distributed load

A fixed-ended beam of uniform section is ten metres long and subject to a uniformly distributed load applied over the entire length of the beam. The elastic moment of resistance of the cross-section of the beam is 100 kN m and the shape factor is 1.10.

Find the maximum value of the load which may be applied without exceeding elastic conditions and compare it with the value obtainable immediately prior to total plastic collapse.

(UL)

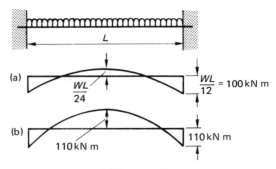

Figure 9.3

Solution The bending moment diagram for the elastic condition is shown in Figure 9.3(a). As shown in Volume 1, Chapter 7 the maximum value occurs at the supports and equals $WL/12$. With the value given in the question

$WL/12 = 100$ kN m

$W = 100$ kN m $\times 12/10$ m

$\quad = 120$ kN total or 12 kN/m   (*Ans*)

The shape factor for the section is 1.10 and therefore a plastic hinge is formed when the bending moment reaches 110 kN m. This occurs first at the ends but the beam can withstand further loading because it then behaves as if it were simply supported. The bending moment at the ends remains at 110 kN m but that at mid-span increases until it reaches the same value. At this stage a plastic hinge is formed at mid-span and the beam can withstand no further load. The bending moment diagram for this condition is shown in Figure 9.3(b). By symmetry the end reactions are each $\frac{1}{2}W$ and, taking moments at mid-span,

$110 = \frac{1}{2}W \times \frac{1}{2}L - \frac{1}{2}W \times \frac{1}{4}L - 110$

from which $WL/8 = 220$, and

$W = 220 \times 8/10 = 176$ kN total or 17.6 kN/m   (*Ans*)

Hence the ratio required is

$$\frac{\text{Maximum load under elastic conditions}}{\text{Load at point of plastic collapse}} = \frac{120}{176} = \frac{15}{22} \qquad (Ans)$$

9.3 Plastic bending of T-section beam

A beam has a T cross-section and is made by welding together two plates, the cross-sections of which are rectangular, with dimensions 160 mm $\times$ 10 mm. The beam is subjected to a bending moment of such magnitude that yielding occurs over the lower 40 mm of the web although the yield stress is not developed at the top surface of the flange. The yield stress of $250 \, MN/m^2$ may be assumed to be constant over the area which has yielded while over the remainder of the section the stress is proportional to the distance from the neutral axis. Under these conditions determine the position of the neutral axis and the moment of resistance of the section.

Compare the moment of resistance with that obtained when the section is fully plastic. (UL)

Solution The cross-section is shown in Figure 9.4(a) and the distribution of stress is of the form shown in Figure 9.4(b). The neutral axis does not pass through the centroid of the section. Let its distance from the top edge of the flange be x mm. Then by proportion,

$$\text{Stress at top of flange} = \left(\frac{x}{130 - x}\right)\sigma_Y$$

$$\text{Stress at bottom of flange} = \left(\frac{x - 10}{130 - x}\right)\sigma_Y$$

$$\text{Average stress in flange} = \left(\frac{x - 5}{130 - x}\right)\sigma_Y$$

Working in millimetres, the forces in the four parts of the cross-section are as follows:

Flange:

$$\left(\frac{x - 5}{130 - x}\right)\sigma_Y \times 160 \times 10 = 1\,600\sigma_Y\left(\frac{x - 5}{130 - x}\right)$$

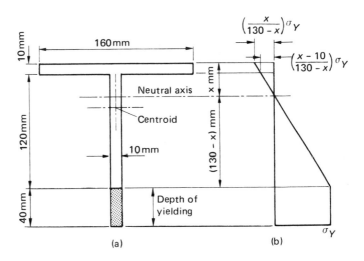

Figure 9.4

Web above n.a.:

$$\frac{1}{2}\left(\frac{x-10}{130-x}\right)\sigma_Y \times (x-10) \times 10 = 5\sigma_Y\frac{(x-10)^2}{(130-x)}$$

Elastic portion of web below n.a.:

$$\tfrac{1}{2}\sigma_Y \times (130-x) \times 10 = 5\sigma_Y(130-x)$$

Plastic portion of web:

$$\sigma_Y \times 40 \times 10 = 400\sigma_Y$$

Since the resultant force on the section is zero, the total forces above and below the n.a. are equal. Hence:

$$1\,600\sigma_Y\left(\frac{x-5}{130-x}\right) + 5\sigma_Y\frac{(x-10)^2}{(130-x)} = 5\sigma_Y(130-x) + 400\sigma_Y$$

Multiplying through by $(130-x)/5\sigma_Y$ and collecting terms,

$$640x = 28\,800 \qquad x = 45$$

The neutral axis is therefore 45 mm from the top edge. (For comparison, the centroid is 47.5 mm from the top edge.) (*Ans*)

With this result the forces for the four portions of the cross-section become

Flange:

$$(1\,600 \times 10^{-6}\,\text{m}^2) \times (250 \times 10^6\,\text{N/m}^2) \times \left(\frac{45-5}{130-45}\right) = 188.2\,\text{kN}$$

Web above n.a.:

$$(5 \times 10^{-3}\,\text{m}) \times (250 \times 10^6\,\text{N/m}^2) \times \frac{(45-10)^2}{(130-45)} \times 10^{-3}\,\text{m} = 18.0\,\text{kN}$$

Elastic portion of web below n.a.:

$$(5 \times 10^{-3}\,\text{m}) \times (250 \times 10^6\,\text{N/m}^2) \times (130-45) \times 10^{-3}\,\text{m} = 106.3\,\text{kN}$$

Plastic portion of web:

$$(400 \times 10^{-6}\,\text{m}^2) \times (250 \times 10^6\,\text{N/m}^2) = 100.0\,\text{kN}$$

The centroid of a trapezium whose parallel sides a and b are distance h apart, is

$$(a + 2b)h/3(a + b)$$

from the side a and hence the distance from the n.a. at which the flange force effectively acts is 40.2 mm.

The corresponding distances for the other three forces are 23.3 mm, 56.7 mm and 105 mm respectively. The moment of resistance of the section is, taking moments about the n.a.,

$$M = (188.2 \times 10^3 \text{ N}) \times (40.2 \times 10^{-3} \text{ m})$$
$$+ (18.0 \times 10^3 \text{ N}) \times (23.3 \times 10^{-3} \text{ m})$$
$$+ (106.3 \times 10^3 \text{ N}) \times (57.7 \times 10^{-3} \text{ m})$$
$$+ (100 \times 10^3 \text{ N}) \times (105 \times 10^{-3} \text{ m})$$
$$= 24.6 \text{ kN m} \qquad\qquad (Ans)$$

When the section is fully plastic the areas above and below the n.a. are equal since the resultant force is zero. In the present case the flange and web are equal in area and the n.a. is located at the junction between them.

The forces in the flange and web are each ($\sigma_Y \times$ area), i.e.

$$(250 \times 10^6 \text{ N/m}^2) \times (160 \times 10 \times 10^{-6} \text{ m}^2) = 400 \text{ kN}$$

The distance between them is 85 mm. The moment of resistance is therefore

$$M = (400 \times 10^3 \text{ N}) \times (85 \times 10^{-3} \text{ m}) = 34 \text{ kN m}$$

The ratio of the moments of resistance for the two conditions is therefore $24.6/34.0 = 0.724$ $\qquad\qquad (Ans)$

9.4 Plastic deformation in torsion of solid circular shaft

A solid shaft 1 m long and 40 mm diameter is made of a material with a yield stress in shear of 150 MN/m². The elastic modulus of rigidity is 90 GN/m². Determine

(a) The total angle of twist in radians and the torque in kN m when the material of the shaft just reaches its yield stress, and
(b) the torque in kN m required to increase the angle of twist to twice that at yield.

Prove any formulae used. $\qquad\qquad (UL)$

Solution (a) The stress distribution for the elastic condition is shown in Figure 9.5(a). The proof of the torsion equation for this case is

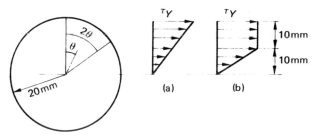

Figure 9.5

given in Volume 1, Chapter 5 where it is shown that

$$\tau/r = T/J = G\theta/L$$

and, with the figures given in the equation,

$$\theta = \tau L/Gr$$

$$= \frac{(150 \times 10^6 \, \text{N/m}^2) \times (1 \, \text{m})}{(90 \times 10^9 \, \text{N/m}^2) \times (20 \times 10^{-3} \, \text{m})}$$

$$= \tfrac{1}{12} \, \text{rad} \qquad\qquad (Ans)$$

The polar second moment

$$J = \pi d^4/32 = \pi \times (40 \times 10^{-3} \, \text{m})^4/32$$

$$= 2.51 \times 10^{-7} \, \text{m}^4$$

Hence the torque is

$$T = \tau J/r = \frac{(150 \times 10^6 \, \text{N/m}^2) \times (2.51 \times 10^{-7} \, \text{m}^4)}{20 \times 10^{-3} \, \text{m}}$$

$$= 1.88 \, \text{kN m} \qquad\qquad (Ans)$$

(b) If the angle of twist is doubled the strain at each radius of the cross-section is doubled and the corresponding distribution of stress is shown in Figure 9.5(b). The central part of the shaft, up to a radius of 10 mm, remains elastic and the usual theory can be applied.

$$J = \tfrac{1}{32}\pi \times (20 \times 10^{-3} \, \text{m})^4 = 1.57 \times 10^{-8} \, \text{m}^4$$

and the elastic torque is

$$T = \tau J/r = \frac{(150 \times 10^6 \, \text{N/m}^2) \times (1.57 \times 10^{-8} \, \text{m}^4)}{10 \times 10^{-3} \, \text{m}}$$

$$= 0.235 \, \text{kN m}$$

For the outer part of the shaft where yielding occurs the stress is constant at the yield value τ_Y. For an annular ring at radius r, radial thickness δr, the torque δT is

$$\delta T = \text{stress} \times \text{area} \times \text{radius}$$

$$= \tau_Y \times 2\pi r \delta r \times r$$

Integrating between radii of 10 mm and 20 mm the total torque is

$$T = \int_{0.01}^{0.02} 2\pi r^2 \tau_Y \, dr = 2\pi \times 150 \times 10^6 \left[\frac{r^3}{3}\right]_{0.01}^{0.02}$$

$$= 700\pi \, \text{N m} = 2.20 \, \text{kN m}$$

Hence the torque required to increase the angle of twist to twice that at yield is $0.235 + 2.20 = 2.44 \, \text{kN n}$. $\qquad (Ans)$

A cantilever consists of an 80 mm × 80 mm × 10 mm angle with the top face AB horizontal (Figure 9.6). It carries a load of 2 kN at a distance of 1 m from the fixed end, the line of action of the load passing through the centroid of the section and inclined at 30° to the vertical.

Determine the stress at the corners A, B and C at the fixed end and also the position of the neutral axis.   (*UL*)

Solution The theory of simple bending (Volume 1, Chapter 3) assumes that the cross-section of the beam is symmetrical about the plane of bending and that the loads are applied in this plane. In the present example the principal axes of the cross-section are UU (an axis of symmetry) and VV which is perpendicular to it. These axes make angles of 45° with XX and YY.

The component of the load in the plane UU is 2 kN × cos 15° = 2 kN × 0.966 = 1.932 kN and the corresponding bending moment at the fixed end is $M = 1.932$ kN × 1 m = 1.932 kN m.

The distance OA = 2.34 cm × sec 45° = 2.34 × 1.414 = 3.31 cm and, considering bending about VV the stress at A is

$$\sigma = \frac{M \times OA}{I_{VV}} = \frac{(1.932 \times 10^3 \,\text{N m}) \times (3.31 \times 10^{-2}\,\text{m})}{36.3 \times 10^{-8}\,\text{m}^4}$$

$$= 176 \,\text{MN/m}^2 \,(\text{tensile}) \qquad (Ans)$$

There is no stress at A for bending about UU.

The perpendicular distance of B and C from the axis VV is

$$(AB \cos 45° - AO) = (8.0 \,\text{cm} \times 0.707 - 3.31 \,\text{cm}) = 2.35 \,\text{cm}$$

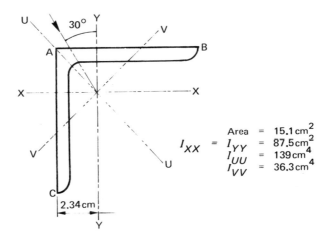

Area = 15.1 cm^2
$I_{XX} = I_{YY}$ = 87.5 cm^2
I_{UU} = 139 cm^4
I_{VV} = 36.3 cm^4

Figure 9.6

and thus, for bending about VV the stress at B and C is

$$\sigma = \frac{(1.932 \times 10^3 \, \text{N m}) \times (2.35 \times 10^{-2} \, \text{m})}{36.3 \times 10^{-8} \, \text{m}^4}$$

$$= 125.0 \, \text{MN/m}^2 \, \text{(compressive)}$$

The component of the load in the plane VV is

$$2 \, \text{kN} \times \cos 75° = 2 \times 0.259 = 0.518 \, \text{kN}$$

and the corresponding bending moment at the fixed end is

$$0.518 \, \text{kN} \times 3 \, \text{m} = 1.554 \, \text{kN m}$$

The perpendicular distance from B (and C) on to UU is

$$AB \cos 45° = 8.0 \, \text{cm} \times 0.707 = 5.66 \, \text{cm}$$

The stress at B and C due to bending about UU is therefore

$$\sigma = \frac{(1.554 \times 10^3 \, \text{N m}) \times (5.66 \times 10^{-2} \, \text{m})}{139 \times 10^{-8} \, \text{m}^4}$$

$$= 63.4 \, \text{MN/m}^2 \, \text{(tensile at B, compressive at C)}$$

Combining these results,

$$\text{stress at B} = 125.0 - 63.4 = 61.6 \, \text{MN/m}^2 \, \text{(compressive)} \qquad (Ans)$$

$$\text{stress at C} = 125.0 + 63.4 = 188.4 \, \text{MN/m}^2 \, \text{(compressive)} \qquad (Ans)$$

The neutral axis passes through O; let α be the angle (measured anticlockwise) which it makes with VV. If P is a point on the neutral axis having coordinates (u, v) relative to the axes VV and UU then $u = OP \sin \alpha$ and $v = OP \cos \alpha$. A moment M applied in the plane of the original load has components $M \cos 15°$ and $M \sin 15°$ about the axes VV and UU respectively. Since the resultant stress at P is zero we have

$$\frac{M \cos 15° \times OP \sin \alpha}{I_{VV}} + \frac{M \sin 15° \times OP \cos \alpha}{I_{UU}} = 0$$

or

$$\tan \alpha = -\frac{I_{VV}}{I_{UU}} \tan 15° = -\frac{36.3}{139} \tan 15° = -0.070 \qquad \alpha = -4° \, 0'$$

The neutral axis is therefore in a direction of 4° 0′ clockwise from VV.
(Ans)

9.6 Derivation of expression for asymmetrical bending stress variation

Taking as a starting point the elementary knowledge that $\sigma/y = E/R$, in pure bending, show that the stress at a point (x, y) in a cross-section subjected to a direct load W and bending moments M_{XX} and M_{YY} about axes XX and YY which are

Solution Suppose (Figure 9.7) the principal axes UU and VV make an angle θ with the given axes XX and YY. Let the point (x, y) have coordinates (u, v) relative to these axes.

The moment M_{XX} has components $M_{XX} \cos \theta$ and $M_{XX} \sin \theta$ about the axes UU and VV respectively and thus the stress at (x, y) due to M_{XX} is

$$\sigma = \frac{M_{XX} \cos \theta \cdot v}{I_{UU}} + \frac{M_{XX} \sin \theta \cdot u}{I_{VV}}$$

As shown in the Appendix, $u = x \cos \theta + y \sin \theta$ and $v = y \cos \theta - x \sin \theta$. Thus the stress becomes

$$\sigma = M_{XX}\left[\frac{\cos \theta(y \cos \theta - x \sin \theta)}{I_{UU}} + \frac{\sin \theta(x \cos \theta + y \sin \theta)}{I_{VV}}\right]$$

$$= \frac{M_{XX}}{I_{UU}I_{VV}}\left[\begin{array}{c}x(I_{UU} \sin \theta \cos \theta - I_{VV} \sin \theta \cos \theta) \\ + y(I_{UU} \sin^2 \theta + I_{VV} \cos^2 \theta)\end{array}\right]$$

$$= \frac{M_{XX}}{I_{UU}I_{VV}}\left[\begin{array}{c}\frac{1}{2}x \sin 2\theta(I_{UU} - I_{VV}) + \frac{1}{2}y(I_{UU} + I_{VV}) \\ + \frac{1}{2}y \cos 2\theta(I_{VV} - I_{UU})\end{array}\right]$$

But, from equations (vi) and (x) of the Appendix,

$$I_{UU} + I_{VV} = I_{XX} + I_{YY} \qquad I_{VV} - I_{UU} = 2I_{XY} \operatorname{cosec} 2\theta$$

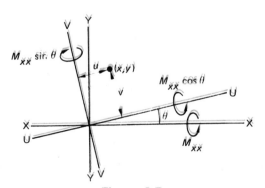

Figure 9.7

Thus:

$$\sigma = \frac{M_{XX}}{I_{UU}I_{VV}}[-xI_{XY} + \tfrac{1}{2}y(I_{XX} + I_{YY}) + yI_{XY}\cot 2\theta]$$

Again, from equations (iii) and (ix) of the Appendix,

$$\cot 2\theta = \frac{I_{YY} - I_{XX}}{2I_{XY}} \qquad I_{UU}I_{VV} = I_{XX}I_{YY} - I_{XY}^2$$

The stress due to M_{XX} is therefore

$$\sigma = \frac{M_{XX}}{I_{XX}I_{YY} - I_{XY}^2} \times \left[-xI_{XY} + \tfrac{1}{2}y(I_{XX} + I_{YY}) + yI_{XY}\left(\frac{I_{YY} - I_{XX}}{2I_{XY}}\right)\right]$$

$$= \frac{M_{XX}(-xI_{XY} + yI_{YY})}{I_{XX}I_{YY} - I_{XY}^2}$$

Similarly the stress due to M_{YY} is

$$\sigma = \frac{M_{YY}(-yI_{XY} + xI_{XX})}{I_{XX}I_{YY} - I_{XX}^2}$$

Adding these results and allowing for the direct stress W/A, the resultant stress is

$$\sigma = \frac{W}{A} + \frac{(M_{XX}I_{YY} - M_{YY}I_{XY})y + (M_{YY}I_{XX} - M_{XX}I_{XY})x}{I_{XX}I_{YY} - I_{XY}^2}$$

as required.

Let ABC be the given angle section as shown in Figure 9.8. The centroid of the section O is $a/4$ from each leg. Using the theorem of parallel axes

$$I_{YY} = I_{XX} = (I \text{ for leg AB})_{XX} + (I \text{ for leg BC})_{XX}$$

$$= at \times (\tfrac{1}{4}a)^2 + \frac{ta^3}{12} + at \times (\tfrac{1}{4}a)^2$$

$$= 5a^3t/24$$

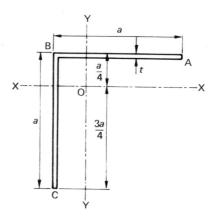

Figure 9.8

In calculating the product moment each leg can be considered as concentrated at its own centroid. Hence:

$$I_{XY} = 2 \times at \times \tfrac{1}{4}a \times \tfrac{1}{4}a = \tfrac{1}{8}a^3 t$$

For the given loading, $M_{XX} = wL^2/8$ and $M_{YY} = 0$. The maximum stress occurs at one of the corners A, B or C. Considering C, $x = -a/4$ and $y = -3a/4$. Using the result derived in the first part of the solution

$$\sigma = \frac{(wL^2/8) \times (5a^3 t/24) \times (-3a/4) + (-wL^2/8) \times (a^3 t/8) \times (-a/4)}{(5a^3 t/24) \times (5a^3 t/24) - (a^3 t/8)^2}$$

$$= \left(\frac{wL^2}{8a^2 t}\right) \frac{(-\tfrac{5}{24} \times \tfrac{3}{4}) + (\tfrac{1}{8} \times \tfrac{1}{4})}{(\tfrac{5}{24})^2 - (\tfrac{1}{8})^2}$$

$$= -9wL^2/16a^2 t$$

The reader should check that the stresses at the other extremities A and B are numerically smaller and this result is therefore the maximum bending stress induced.

9.7 Central deflection of rectangular plate built in along edges due to uniform normal load

A thin flat plate of length L, breadth B, and thickness t is built in along its edges and subjected to a normal loading of w per unit area. From consideration of the central deflection show that the maximum bending stress at the centre will be given approximately by

$$\frac{wB^2 L^4}{4t^2(L^4 + B^4)} \qquad (UL)$$

Solution First consider bending in the plane parallel to the length L as shown in Figure 9.9 and let w_1 be the corresponding part of the normal loading. The deflection and stress due to this loading are the same as for a built-in beam, length L and rectangular cross-section, width B and depth t. The total load $W = w_1 BL$ and, for the cross-section, $I = Bt^3/12$. Thus the central deflection is

$$\delta = \frac{WL^3}{384EI} = \frac{(w_1 BL)L^3 \times 12}{384EBt^3}$$

$$= \frac{w_1 L^4}{32Et^3} \qquad (i)$$

For bending in the plane parallel to the breadth B the corresponding

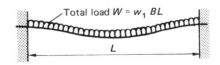
Total load $W = w_1 BL$

L

Figure 9.9

load is $(w - w_1)$ and the central deflection is

$$\delta = \frac{(w - w_1)B^4}{32Et^3} \qquad \text{(ii)}$$

Equating (i) and (ii),

$$w_1 L^4 = (w - w_1)B^4$$

$$w_1 = \frac{wB^4}{L^4 + B^4} \qquad \text{(iii)}$$

The bending moment at mid-span is $WL/24$ (see Volume 1, Chapter 7) and the corresponding maximum stress is

$$\sigma = \frac{My}{I} = \frac{(WL/24)(t/2)}{(Bt^3/12)} = \frac{WL}{4Bt^2}$$

Substituting from (iii)

$$W = w_1 BL = \left(\frac{wB^4}{L^4 + B^4}\right)BL$$

$$\sigma = \frac{wL^2 B^4}{4t^2(L^4 + B^4)}$$

Interchanging B and L, the corresponding stress for bending in the perpendicular plane is

$$\sigma = \frac{wB^2 L^4}{4t^2(L^4 + B^4)}$$

Since L is greater than B this is the required maximum stress.

9.8 Differential equation of bending for circular flat plate supported at rim

A circular flat plate of uniform thickness is supported uniformly round its rim and is subjected on one side to a normal pressure which on any concentric ring is uniformly distributed. Considering a diametral section of the plate derive the differential equation

$$\frac{d^2\theta}{dr^2} + \frac{1}{r}\frac{d\theta}{dr} - \frac{\theta}{r^2} + \frac{S}{D} = 0$$

where θ = slope at any radius r,
S = shearing force per unit length of arc at radius r,
D = flexural rigidity of plate
$= Et^3/12(1 - v^2)$
where t = thickness of plate and v = Poisson's ratio. *(UL)*

Solution Take the origin O at the centre of the deflected plate as shown in Figure 9.10 with axes OX and OY in the plane of a diametral

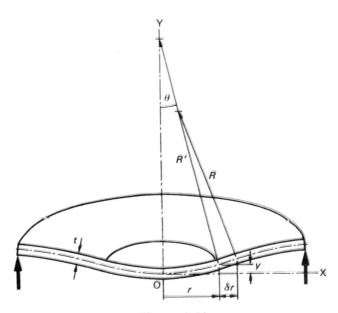

Figure 9.10

section. Let y be the deflection at radius r so that the slope in this plane $\theta = dy/dr$. Bending occurs not only in the XOY plane but also in a plane at right angles to it. Let the radii of curvature in these planes (at radius r) be R and R'. Then, as shown in Volume 1, Chapter 6,

$$\frac{1}{R} = \frac{d^2 y}{dx^2} = \frac{d\theta}{dr} \tag{i}$$

Since the loading, and hence the distortion, are symmetrical a circumferential 'section', originally cylindrical, becomes part of a cone whose apex is on the axis OY. Hence the radius of curvature R' is given by $r = R'\theta$ and

$$1/R' = \theta/r \tag{ii}$$

Using the relationship between strain and radius of curvature (see Volume 1, Chapter 3) and equations (i) and (ii) we have, at the surface of the plate,

$$\varepsilon = \frac{\frac{1}{2}t}{R} = \frac{1}{2}t\frac{d\theta}{dr}$$

$$\varepsilon' = \frac{\frac{1}{2}t}{R'} = \frac{1}{2}t\frac{\theta}{r}$$

The stresses in the same directions at the same point (see Chapter 2) are

$$\sigma = \frac{E}{1 - v^2}(\varepsilon + v\varepsilon') = \frac{tE}{2(1 - v^2)}\left(\frac{d\theta}{dr} + \frac{v\theta}{r}\right) \tag{iii}$$

SPECIAL BENDING AND TORSION PROBLEMS **223**

$$\sigma' = \frac{E}{1 - v^2}(\varepsilon' + v\varepsilon) = \frac{tE}{2(1 - v^2)}\left(\frac{\theta}{r} + v\frac{d\theta}{dr}\right) \tag{iv}$$

Consider the bending of an element of unit length, the maximum stress being σ. The distance from the neutral axis to the surface is $t/2$, $I = (1 \times t^3)/12$ and, from the bending equation $\sigma/y = M/I$,

$$M = \frac{\sigma \times \frac{1}{12}(1 \times t^3)}{\frac{1}{2}t} = \frac{1}{6}\sigma t^2$$

Hence, using (iii) and (iv)

$$M = \frac{t^3 E}{12(1 - v^2)}\left(\frac{d\theta}{dr} + \frac{v\theta}{r}\right) = D\left(\frac{d\theta}{dr} + \frac{v\theta}{r}\right) \tag{v}$$

$$M' = \frac{t^3 E}{12(1 - v^2)}\left(\frac{\theta}{r} + v\frac{d\theta}{dr}\right) = D\left(\frac{\theta}{r} + v\frac{d\theta}{dr}\right) \tag{vi}$$

where $D = t^3 E/12(1 - v^2)$, the *flexural rigidity*.

Figure 9.11 shows a small element of the plate which subtends a small angle ϕ at the centre. M and $M + \delta M$ are the moments per unit length at radii r and $r + \delta r$ in the diametral plane, S and $S + \delta S$ are the corresponding shearing forces per unit length, M' is the moment per unit length in the perpendicular plane and the component of the moment $M'\delta r$ in the central plane of the element through the centre is

$$M'\delta r \cos\left(90° - \tfrac{1}{2}\phi\right) = M'\delta r \sin\tfrac{1}{2}\phi = \tfrac{1}{2}M'\delta r\phi$$

since ϕ is a small angle.

Taking moments in this plane about a point at radius $(r + \delta r)$ we have

$$(M + \delta M)\phi(r + \delta r) - M\phi r - \tfrac{1}{2}(2M'\delta r\phi) + S\phi r\delta r = 0$$

Cancelling through by ϕ and neglecting products of small quantities, we obtain

$$M\delta r + r\delta M - M'\delta r + Sr\delta r = 0$$

or dividing by δr and taking the limiting case

$$M + r\frac{dM}{dr} - M' + Sr = 0$$

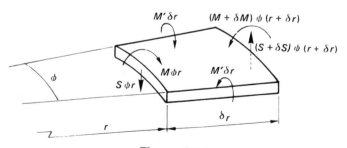

Figure 9.11

Substituting from (v) and (vi),

$$D\left(\frac{d\theta}{dr} + \frac{v\theta}{r}\right) + rD\left[\frac{d^2\theta}{dr^2} + v\left(\frac{r\,d\theta/dr - \theta}{r^2}\right)\right] - D\left(\frac{\theta}{r} + v\frac{d\theta}{dr}\right) + Sr = 0$$

or, dividing through by rD and rearranging,

$$\frac{d^2\theta}{dr^2} + \frac{1}{r}\frac{d\theta}{dr} - \frac{\theta}{r^2} + \frac{S}{D} = 0$$

as required.

9.9 Stress and deflection of circular diaphragm, supported at edge, due to uniform pressure

A diaphragm of uniform thickness is 10 cm diameter. It is freely supported at its edge and loaded on one face with a uniform pressure of 20 kN/m².

If the maximum stress is limited to 100 MN/m² find the necessary thickness of the diaphragm and determine also the maximum deflection produced.

At any radius r, the bending moment per unit length of arc is

$$M = D\left(\frac{d\theta}{dr} + v\frac{\theta}{r}\right)$$

where θ is the slope at radius r.

$$\frac{d}{dr}\left[\frac{1}{r}\frac{d(r\theta)}{dr}\right] + \frac{S}{D} = 0$$

where S = the shearing force per unit length of arc at radius r
and D = the flexural rigidity of the plate
$$= Et^3/12(1 - v^2)$$

$E = 200\ \text{GN/m}^2;\ v = 0.28$ (UL)

Solution The second equation given in the question is equivalent to the relationship derived in the previous solution. Working in symbols, let w per unit area be the uniform pressure. Then at radius r, the total shear force around the circumference equals the total load within the circumference and

$$2\pi rS = w\pi r^2$$
$$S = \tfrac{1}{2}wr$$

Substituting in the given equation

$$\frac{d}{dr}\left[\frac{1}{r}\frac{d(r\theta)}{dr}\right] = -\frac{wr}{2D}$$

and, on integrating once,

$$\frac{1}{r}\frac{d(r\theta)}{dr} = -\frac{wr^2}{4D} + A$$

$$\frac{d(r\theta)}{dr} = -\frac{wr^3}{4D} + Ar$$

where A is a constant. On integrating again,

$$r\theta = -\frac{wr^4}{16D} + \tfrac{1}{2}Ar^2 + B$$

(i)

$$\theta = -\frac{wr^3}{16D} + \tfrac{1}{2}Ar + \frac{B}{r}$$

where B is a second constant. But $\theta = dy/dr$ and at $r = 0$ the slope $= 0$. Hence B must be zero and

$$\frac{dy}{dr} = -\frac{wr^3}{16D} + \tfrac{1}{2}Ar$$

(ii)

from which

$$y = -\frac{wr^4}{64D} + \tfrac{1}{4}Ar^2 + C$$

(iii)

where C is a third constant of integration.

But $y = 0$ when $r = 0$ and therefore $C = 0$. From (i),

$$\frac{d\theta}{dr} = -\frac{3wr^2}{16D} + \tfrac{1}{2}A$$

and, substituting in the expression for M given in the question

$$M = D\left(-\frac{3wr^2}{16D} + \tfrac{1}{2}A - \frac{vwr^2}{16D} + \tfrac{1}{2}vA\right)$$

$$= \tfrac{1}{2}AD(1 + v) - \tfrac{1}{16}wr^2(3 + v)$$

(iv)

Since the plate is freely supported at its edge $M = 0$ when $r = R$, the external radius. Thus putting $M = 0$ and $r = R$ in (iv),

$$\tfrac{1}{2}AD(1 + v) = \tfrac{1}{16}wR^2(3 + v)$$

$$A = \frac{wR^2}{8D}\left(\frac{3 + v}{1 + v}\right)$$

(v)

Using this value of A and putting $r = 0$ for maximum bending moment we obtain, from (iv),

$$M_{\text{max}} = \tfrac{1}{16}wR^2(3 + v)$$

This result is the bending moment per unit length so putting $I = \tfrac{1}{12}(1 \times t^3)$ and y (distance from neutral axis) $= \tfrac{1}{2}t$ the maximum stress is

$$\sigma = \frac{My}{I} = \frac{\tfrac{1}{16}wR^2(3 + v) \times \tfrac{1}{2}t}{\tfrac{1}{12}(1 \times t^3)} = \frac{3wR^2(3 + v)}{8t^2}$$

Rearranging and substituting the numerical data given in the question,

$$t^2 = \frac{3wR^2(3+v)}{8\sigma}$$

$$= \frac{3 \times (20 \times 10^3 \text{ N/m}^2) \times (0.05 \text{ m})^2 \times (3+0.28)}{8 \times (100 \times 10^6 \text{ N/m}^2)}$$

$$= 6.15 \times 10^{-7} \text{ m}^2$$

The necessary thickness t is therefore

$$7.84 \times 10^{-4} \text{ m} \quad \text{or} \quad 0.784 \text{ mm} \qquad (Ans)$$

Since the origin of coordinates is taken at the centre of the *deflected* plate the maximum deflection is obtained by putting $r = R$. Hence, from (iii), with the value of A given by (v)

$$y_{max} = -\frac{wR^4}{64D} + \frac{wR^2}{8D}\left(\frac{3+v}{1+v}\right) \times \frac{R^2}{4}$$

$$= \frac{wR^4}{64D}\left(\frac{5+v}{1+v}\right)$$

Substituting the figures given in the question

$$D = \frac{Et^3}{12(1-v^2)} = \frac{(200 \times 10^9 \text{ N/m}^2) \times (7.84 \times 10^{-4} \text{ m})^3}{12(1-0.28^2)}$$

$$= 8.71 \text{ N m}$$

and the maximum deflection

$$y_{max} = \frac{wR^4}{64D}\left(\frac{5+v}{1+v}\right) = \frac{(20 \times 10^3 \text{ N/m}^2) \times (0.05 \text{ m})^4}{64 \times (8.71 \text{ N m})} \times \left(\frac{5+0.28}{1+0.28}\right)$$

$$= 0.925 \times 10^{-3} \text{ m} \quad \text{or} \quad 0.925 \text{ mm} \qquad (Ans)$$

9.10 Radial and circumferential stresses in circular plate, clamped at edge, due to central point load

A uniform circular plate of radius a and thickness t is clamped along its edge in such a manner that the slope and deflection of the plate at the edge are zero. The plate carries a downward central load P on the top surface. The strains at the lower surface of the plate, at radius r, in the radial and circumferential directions are given by

$$\varepsilon_r = \tfrac{1}{2}t\frac{d\phi}{dr} \quad \text{and} \quad \varepsilon_\theta = \frac{t\phi}{2r} \quad \text{respectively}$$

where ϕ is the slope of the surface of the plate in the radial direction.

With Young's modulus $= E$ and Poisson's ratio $= v$ obtain expressions for the stresses, σ_r and σ_θ, at the lower surface of the plate, in the radial and circumferential directions respectively.

Obtain a general expression for the deflection of the plate, and show that the maximum deflection is given by

$$Pa^2/16\pi D$$

where D is the flexural rigidity.

The following relationship may be used:

$$\frac{d}{dr}\left[\frac{1}{r}\frac{d}{dr}(r\phi)\right] = \pm\frac{Q}{D}$$

where Q is the shear force/circumferential length at radius r (the sign in front of Q/D depends on the sign convention adopted).

Show that as $r \to 0$ so $\sigma_r \to \infty$ and $\sigma_\theta \to \infty$. Comment on the fact that the maximum deflection is finite although the stresses near the centre approach infinity.   (UL)

Solution   At radius r the total shear force on a circumferential ring is P the central load. Hence the shear force per circumferential length is $Q = P/2\pi r$ and substituting in the differential equation

$$\frac{d}{dr}\left[\frac{1}{r}\frac{d}{dr}(r\phi)\right] = -\frac{Q}{D} = -\frac{P}{2\pi rD}$$

Integrating,

$$\frac{1}{r}\frac{d}{dr}(r\phi) = -\frac{P}{2\pi D}\ln r + A$$

$$\frac{d}{dr}(r\phi) = -\frac{Pr}{2\pi D}\ln r + Ar$$

where A is a constant.

Integrating again and noting that the integral of $r \ln r$ is $\frac{1}{2}r^2(\ln r - \frac{1}{2})$, we have

$$r\phi = -\frac{Pr^2}{4\pi D}(\ln r - \frac{1}{2}) + \frac{1}{2}Ar^2 + B$$

where B is a second constant. At $r = 0$, the slope $\phi = 0$ and $r^2 \ln r$ tends to zero. Thus $B = 0$ and

$$\phi = -\frac{Pr}{4\pi D}(\ln r - \frac{1}{2}) + \frac{1}{2}Ar \tag{i}$$

Also, since the edge is clamped, $\phi = 0$ when $r = a$ and

$$A = \frac{P}{2\pi D}(\ln a - \frac{1}{2})$$

Substituting this result in (i), the slope is

$$\frac{dy}{dr} = \phi = -\frac{Pr}{4\pi D}(\ln r - \frac{1}{2}) + \frac{Pr}{4\pi D}(\ln a - \frac{1}{2}) = \frac{Pr}{4\pi D}(\ln a - \ln r) \tag{ii}$$

Integrating to obtain the deflection we have

$$y = \frac{Pr^2}{8\pi D} \ln a - \frac{Pr^2}{8\pi D}(\ln r - \tfrac{1}{2}) + C$$

where C is a further constant. If the origin is taken at the centre of the *deflected* plate, $y = 0$ when $r = 0$ and hence $C = 0$. Thus the deflection equation is

$$y = \frac{Pr^2}{8\pi D}(\ln a - \ln r + \tfrac{1}{2})$$

$$= \frac{Pr^2}{8\pi D}\left(\ln \frac{a}{r} + \frac{1}{2}\right)$$

Putting $r = a$, the maximum deflection is

$$y_{max} = \frac{Pa^2}{8\pi D}\left(\ln \frac{a}{a} + \frac{1}{2}\right) = \frac{Pa^2}{16\pi D}$$

as required.

From (ii), $\phi = \frac{Pr}{4\pi D} \ln \frac{a}{r}$ and by differentiation

$$\frac{d\phi}{dr} = \frac{P}{4\pi D}\left(\ln \frac{a}{r} - 1\right)$$

Substituting in the expressions given in the question the strains are

$$\varepsilon_r = \frac{Pt}{8\pi D}\left(\ln \frac{a}{r} - 1\right) \quad \text{and} \quad \varepsilon_\theta = \frac{Pt}{8\pi D} \ln \frac{a}{r}$$

The corresponding stresses are given by:

$$\sigma_r = \frac{E}{1 - v^2}(\varepsilon_r + v\varepsilon_\theta)$$

$$= \left(\frac{E}{1 - v^2}\right)\left(\frac{Pt}{8\pi D}\right)\left(\ln \frac{a}{r} - 1 + v \ln \frac{a}{r}\right)$$

which, since $D = Et^3/12(1 - v^2)$, becomes

$$\sigma_r = \frac{3P}{2\pi t^2}\left[(1 + v) \ln \frac{a}{r} - 1\right]$$

and, similarly,

$$\sigma_\theta = \frac{3P}{2\pi t^2}\left[(1 + v) \ln \frac{a}{r} - v\right]$$

From the form of these expressions it can be seen that as $r \to 0$, $\ln a/r \to \infty$ and the stresses tend to infinity.

In practice the load P cannot be applied at a point and more elaborate analysis (see Problem 23, p. 242) allowing for P being distributed over a small area, would lead to finite stresses. That the maximum deflection is finite arises from the fact that deflection is the integral sum of the strains throughout the plate and this can remain finite even though the strains become infinite at one point.

9.11 Maximum bending stress in heavily curved beam of rectangular section

A beam of uniform cross-section has an initial mean radius of curvature which is of the same order as the radial depth of the section. Assuming plane sections remain plane after bending prove that for a pure bending couple the distance of the neutral axis from the centre of curvature is

$$R_n = A \bigg/ \int_{r_1}^{r_2} \frac{dA}{R}$$

where A is the area of the section, dA is the area of an elementary strip at radius R, and r_1, r_2 are the inner and outer radii of the section respectively.

A steel hook of rectangular section has an inner radius of 50 mm and an outer radius of 100 mm. The width is 25 mm. A bending moment of 1 kN m is applied to the section, tending to open the hook. Calculate the maximum tensile and compressive stresses. (UL)

Solution In deriving the theory of simple bending (Volume 1, Chapter 3) it was assumed that the beam was initially straight. The results obtained can be applied to curved beams provided that the initial radius of curvature is large compared with the cross-sectional dimensions.

For heavily curved beams, however, a new analysis is required. In particular the neutral axis no longer passes through the centroid of the cross-section nor is the stress proportional to the distance from the neutral axis.

Suppose, in Figure 9.12(a), P and Q represent the intersections of the neutral plane with two cross-sections which initially intersect at O. Let O' be the intersection of these planes after bending. Suppose R_n and R' are the radii of curvature of the neutral plane before and after bending. Then

$$PQ = R_n\theta = R'(\theta + \delta\theta) \tag{i}$$

If δA is an element of area, distance y below the neutral axis, Figure 9.12(b), the longitudinal strain on the element is

$$\frac{\text{increase in length}}{\text{original length}} = \frac{(R' + y)(\theta + \delta\theta) - (R_n + y)\theta}{(R_n + y)\theta}$$

$$= \frac{R'(\theta + \delta\theta) + y\theta + y\delta\theta - R_n\theta - y\theta}{(R_n + y)\theta}$$

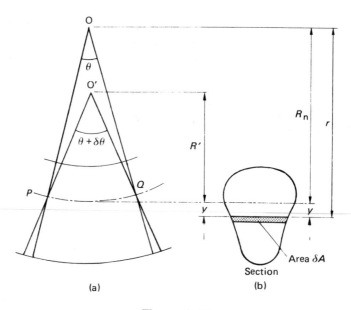

Figure 9.12

and, using (i) this becomes

$$\text{strain} = \frac{y\delta\theta}{(R_n + y)\theta}$$

and the corresponding stress

$$\sigma = \frac{Ey\delta\theta}{(R_n + y)\theta} \tag{ii}$$

Integrating over the whole cross-section,

$$\text{total force} = \frac{E\delta\theta}{\theta}\int_{r_1}^{r_2}\frac{y}{R_n + y}\,dA = \frac{E\delta\theta}{\theta}\int_{r_1}^{r_2}\frac{r - R_n}{r}\,dA \tag{iii}$$

where r is the radius from the original centre of curvature O to the element δA. In pure bending the net force on a cross-section is zero and the integrals in (iii) are therefore zero. Thus

$$\int_{r_1}^{r_2}\left(1 - \frac{R_n}{r}\right)dA = 0$$

$$\int_{r_1}^{r_2}dA = \int_{r_1}^{r_2}\frac{R_n}{r}\,dA$$

The integral on the left is the total area A, R_n is a constant and thus

$$R_n = A\bigg/\int_{r_1}^{r_2}\frac{dA}{r}$$

as required.

Similarly the total moment about the neutral axis is

$$M = \frac{E\delta\theta}{\theta} \int_{r_1}^{r_2} \frac{y^2}{R_n + y} \, dA \qquad (iv)$$

and this integral may be written in two parts thus:

$$\int_{r_1}^{r_2} \frac{(y^2 + R_n y) - R_n y}{R_n + y} \, dA = \int_{r_1}^{r_2} y \, dA - R_n \int_{r_1}^{r_2} \frac{y}{R_n + y} \, dA$$

The first of these integrals is the first moment of the cross-section about the neutral axis. It can be written as Ae where e is the distance of the centroid from the neutral axis. As shown by (iii) the second integral is proportional to the net force on the section and is therefore zero. Equation (iv) therefore becomes

$$M = \frac{E\delta\theta}{\theta} \times Ae$$

and from (ii)

$$E\frac{\delta\theta}{\theta} = \sigma(R_n + y)/y$$

$$M = \frac{\sigma(R_n + y)}{y} Ae \qquad (v)$$

$$\frac{\sigma}{y} = \frac{M}{Ae(R_n + y)}$$

Working in mm units, we have, for the numerical part of the question:

$$A = 50 \times 25 = 1\,250, \qquad r_1 = 50, \qquad r_2 = 100$$

and

$$\delta A = 25 \times \delta r$$

Hence

$$R_n = \frac{A}{\int_{r_1}^{r_2} \dfrac{dA}{r}} = \frac{1\,250}{\int_{50}^{100} \dfrac{25}{r} \, dr}$$

$$= \frac{50}{[\ln r]_{50}^{100}}$$

$$= \frac{50}{\ln 2} = \frac{50}{0.693\,15} = 72.13 \text{ (mm)}$$

The radius of the central axis of the section is 75 mm and hence the neutral axis is 2.87 mm above the centroid in the sense of Figure 9.13. Since this result is obtained as the small difference of two large quantities, R_n must be calculated with appropriate accuracy.

At the inner surface $y = -22.13$ mm and, with $M = -1\,000$ kN mm (M is in the opposite sense to that assumed in the theory) the stress

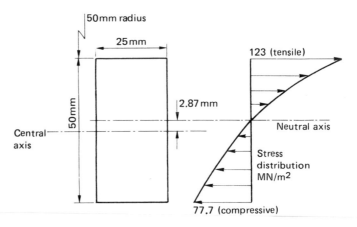

Figure 9.13

there, from (v), is

$$\sigma = \frac{My}{Ae(R_n + y)} = \frac{-1\,000 \times 10^3 \text{ N mm} \times (-22.13 \text{ mm})}{1\,250 \text{ mm}^2 \times 2.87 \text{ mm} \times (72.13 - 22.13) \text{ mm}}$$

$$= 123 \text{ N/mm}^2 \quad \text{or} \quad 123 \text{ MN/m}^2 \text{ (tensile)}$$

Similarly at the outer surface $y = 27.87$ mm and

$$\sigma = -77.7 \text{ N/mm}^2 \quad \text{or} \quad 77.7 \text{ MN/m}^2 \text{ (compressive)} \qquad (Ans)$$

Figure 9.13 shows the variation of stress across the section.

9.12 Location of neutral axis and magnitude of maximum stress in heavily curved beam of rectangular section

Show that, for a curved bar of rectangular cross-section, the distance of the neutral axis from the centroid of the cross-section expressed as a first approximation is

$$e = R\left[1 - \frac{1}{1 + \frac{1}{3}(d/2R)^2}\right]$$

where R is the radius of curvature of the centre line of the bar and d is the radial depth of the cross-section.

A crane hook of rectangular cross-section carries a bending moment of 2 kN m; the radius of the centre line of the hook is 16 cm, the depth of the cross-section is 8 cm and the breadth is 4 cm.

Determine the magnitude of the maximum stress in the cross-section and indicate where it occurs. (UL)

Solution In the notation of the previous example $r_1 = R - \frac{1}{2}d$ and $r_2 = R + \frac{1}{2}d$. Hence the distance of the neutral axis from the original centre of curvature is

$$R_n = \frac{bd}{\int_{R-\frac{1}{2}d}^{R+\frac{1}{2}d} \frac{b}{r}\, dr}$$

$$= \frac{d}{\ln \dfrac{R+\frac{1}{2}d}{R-\frac{1}{2}d}} = \frac{d}{\ln \dfrac{1+(d/2R)}{1-(d/2R)}}$$

From the expansion of $\ln(1+x)$ we obtain

$$\ln\left(\frac{1+x}{1-x}\right) = (x - \tfrac{1}{2}x^2 + \tfrac{1}{3}x^3 - \tfrac{1}{4}x^4 \dots)$$

$$- (-x - \tfrac{1}{2}x^2 - \tfrac{1}{3}x^3 - \tfrac{1}{4}x^4 \dots)$$

$$= 2(x + \tfrac{1}{3}x^3 + \tfrac{1}{5}x^5 \dots)$$

Putting $x = d/2R$ the position of the neutral axis is given by

$$R_n = \frac{d}{2\left[\dfrac{d}{2R} + \dfrac{1}{3}\left(\dfrac{d}{2R}\right)^3 + \dfrac{1}{5}\left(\dfrac{d}{2R}\right)^5 \dots\right]}$$

and the distance between the neutral axis and central axis is

$$e = R - R_n$$

$$= R\left\{1 - \frac{d}{2R\left[\dfrac{d}{2R} + \dfrac{1}{3}\left(\dfrac{d}{2R}\right)^3 + \dfrac{1}{5}\left(\dfrac{d}{2R}\right)^5 + \cdots\right]}\right\}$$

$$= R\left\{1 - \frac{1}{1 + \dfrac{1}{3}\left(\dfrac{d}{2R}\right)^2 + \dfrac{1}{5}\left(\dfrac{d}{2R}\right)^4 + \cdots}\right\}$$

$$= R\left\{1 - \frac{1}{1 + \dfrac{1}{3}\left(\dfrac{d}{2R}\right)^2}\right\}$$

as a first approximation.

This result can be developed further if required by expanding the fraction as a series. Thus

$$\left[1 + \frac{1}{3}\left(\frac{d}{2R}\right)^2 + \frac{1}{5}\left(\frac{d}{2R}\right)^4\right]^{-1}$$

$$= 1 - \left[\frac{1}{3}\left(\frac{d}{2R}\right)^2 + \frac{1}{5}\left(\frac{d}{2R}\right)^4\right] + \left[\frac{1}{3}\left(\frac{d}{2R}\right)^2 + \frac{1}{5}\left(\frac{d}{2R}\right)^4\right]^2$$

$$= 1 - \frac{d^2}{12R^2} - \frac{d^4}{80R^4} + \frac{d^4}{144R^4} : \cdots = 1 - \frac{d^2}{12R^2} - \frac{d^4}{180R^4} \cdots$$

$$e = R\left[1 - \left(1 - \frac{d^2}{12R^2} - \frac{d^4}{180R^4} - \cdots\right)\right] = \frac{d^2}{12R}\left(1 + \frac{d^2}{15R^2} + \cdots\right)$$

The approximation $e \simeq d^2/12R$ is sufficient for most examples. Using it in the numerical part of the question

$$e = \frac{d^2}{12R} = \frac{(8 \text{ cm})^2}{12 \times (16 \text{ cm})} = \tfrac{1}{3} \text{ cm}$$

The maximum stress occurs at the inner surface for which $y = -(4 - \tfrac{1}{3})$ $= -3\tfrac{2}{3}$ cm. The radius to the neutral axis is $(16 - \tfrac{1}{3}) = 15\tfrac{2}{3}$ cm and the area $A = 8 \times 4 = 32$ cm^2. Thus, the maximum stress is

$$\sigma = \frac{My}{Ae(R_n + y)}$$
$$= \frac{2 \text{ kN m} \times (-3\tfrac{2}{3}) \text{ cm}}{32 \text{ cm}^2 \times \tfrac{1}{3} \text{ cm} \times 12 \text{ cm}}$$
$$= 57.3 \text{ MN/m}^2 \hspace{4cm} (Ans)$$

Problems

1 A vertically loaded I-section beam with horizontal flanges 80×20 and vertical web 100×20 (dimensions in mm) is of elastic/perfectly-plastic material with yield stress 320 MN/m^2 in tension and compression. Determine:

(a) its shape factor;
(b) the collapse moment of the cross-section;
(c) the maximum total uniformly distributed load the beam would carry at collapse on a span of 2.5 m with each end firmly built in. (*Brunel*)

Answer (a) 1.274; (b) 77.44 kN m; (c) 496 kN.

2 A length of steel strip with a section 20 mm × 4 mm is to be bent through a right angle about an axis parallel to its longer section dimension. The steel has a yield stress of 400 MN/m^2 and fracture of the steel occurs at a strain of 0.08.

What is the minimum radius to which the strip can be bent before fracture occurs, and what bending moment will be required to make the bend to this radius, assuming the section to be fully plastic? (*PCL*)
Answer 25 mm; 32 N m.

3 A circular shaft of outer diameter 24 mm is solid for part of its length and hollow for the remainder, the bore being 12 mm. The material is elastic/perfectly-plastic with a yield stress in shear of 150 MN/m^2. Determine the torques required to produce the yield stress at a diameter of 18 mm in the hollow portion of the shaft. Find also the diameter to which yielding would then have penetrated in the solid portion of the shaft. (*Brunel*)
Answer 452 N m; 21.0 mm.

4 A horizontal beam has a T-shaped cross-section consisting of a horizontal flange 10 cm wide by 1 cm thick and a vertical web 15 cm deep by 1 cm thick (so that the total depth is 16 cm), the flange being uppermost. The beam may be assumed to be of elastic/perfectly-plastic material. It is subjected to a pure bending moment in the vertical plane of such magnitude that yielding is induced in both the web and the flange. The bending moment is increased until the whole of the flange has just become plastic.

Sketch the distribution of bending stress in the section for this condition if the yield stress in both tension and compression is 250 MN/m², and deduce the position of the neutral axis and how far yielding has progressed up the web.

Determine also the magnitude of the applied bending moment and compare it with the collapse moment of the section.

(Brunel)

Answer 12.5 cm from bottom edge; 10 cm; 27.3 kN m; 0.981.

5 A steel bar of rectangular section 8 cm × 3 cm is used as a simply supported beam on a span of 1.2 m and loaded at mid-span. If the yield stress is 275 MN/m² and the long edges of the section are vertical find the load when yielding first occurs.

Assuming that a further increase in load causes yielding to spread in towards the neutral axis with the stress in the yielded part remaining constant at 275 MN/m², determine the load required to cause yielding for a depth of 1 cm at the top and bottom of the section at mid-span, and find the length of beam over which yielding at the top and bottom faces will have occurred.

(UL)

Answer 29.3 kN; 35.75 kN; 0.215 m.

6 A rectangular steel beam AB 25 mm wide by 12 mm deep is placed symmetrically on two knife-edges C and D, 0.5 m apart, and loaded by applying equal weights at the ends A and B. The steel follows a linear stress–strain law ($E = 200$ GN/m²) up to a yield stress of 280 MN/m²; at this constant stress considerable plastic deformation occurs. It may be assumed that the properties of the steel are the same in tension and compression.

Calculate the bending moment on the central part of the beam CD when yielding commences, and the deflection of the centre relative to the supports.

If the loads are increased until yielding penetrates half-way to the neutral axis calculate the new value of the bending moment and the corresponding deflection.

(UL)

Answer 168 N m; 7.29 mm; 231 N m; 14.6 mm.

7 A steel beam of I-section has an overall depth of 250 mm and width of flange is 150 mm; the flanges are 25 mm thick and the web 10 mm thick. The beam is 5 m long and rests on two supports equidistant from the ends and 3 m apart; the beam carries a point load W at each end. Consider the two loads to be gradually increased but kept equal and assume the yield stress of the material to be 270 MN/m². Determine

(a) the magnitude of W when yielding of the material first occurs, stating clearly where the yield stress is reached

(b) the magnitude of W when yielding just extends through the thickness of the flanges, and

(c) the deflection of the beam at mid-span relative to the supports for the two conditions (a) and (b).

State the assumptions made. For elastic material take $E = 200$ GN/m². *(UL)*

Answer (a) 220.3 kN, along the top and bottom surfaces between supports, (b) 245.8 kN, (c) 12.15, 15.19 mm.

8 The T-section shown in Figure 9.14 is subject to a moment M represented vectorially in the figure and acting in the plane of symmetry of the section which puts the edge AB in compression.

Find the value of the fully plastic moment of resistance of this section and compare it with that moment which first produces yield in the outermost fibres. The yield stress for the material of this section in compression and tension is 210 MN/m². *(UL)*

Answer 29.2 kN m; 16.5 kN m.

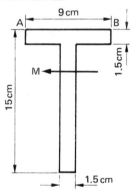

Figure 9.14

9 A steel beam, subjected to a bending moment in the plane YY, has a section symmetrical about YY as shown in Figure 9.15.

The bending moment is such that yielding has occurred in the top portion and the yield stress σ_Y is just reached at the bottom. Assuming that, when yielding has occurred, the stress in the material remains constant at the value σ_Y and that yielding occurs in tension and compression at the same stress, show that the neutral axis is 48.82 mm above the bottom edge and determine the magnitude of the bending moment in terms of σ_Y. *(UL)*

Answer $(101.4\sigma_Y)$ N m where σ_Y is in MN/m².

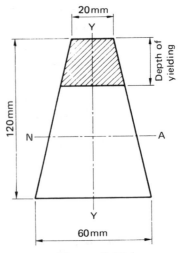

Figure 9.15

10 Making the usual assumptions for the plastic deformation of beams, determine the shape factor for the following sections

(a) circle;
(b) regular hexagon in which one of the diagonals is the neutral axis;
(c) the T-section of Example 9.3.

Answer (a) 1.70, (b) 1.60, (c) 1.81.

11 A rectangular beam 80 mm deep by 40 mm wide is made of a material for which E in compression is 1.5 times E in tension. Find (a) the position of the neutral axis; (b) the moment of resistance when the maximum tensile stress is 88 MN/m^2; (c) the corresponding maximum compressive stress.

If the beam is freely supported on a span of 2.5 m and carries a uniformly distributed load which induces these maximum stresses, find the deflection at the centre of the span. Take E in tension as 60 GN/m^2. (*UL*)
Hint Work from first principles and assume plane sections remain plane.
Answer (a) 4.0 mm from centroidal axis; (b) 4.13 kN m; (c) 108 MN/m^2; 21.7 mm.

12 A steel shaft 8 cm diameter is solid for a certain distance from one end but hollow for the remainder of its length with an inside diameter of 4 cm.

If a pure torque is transmitted from one end of the shaft to the other of such a magnitude that yielding just occurs at the surface of the solid part of the shaft find the depth of yielding in the hollow part of the shaft and the ratio of the angles of twist per unit length for the two parts of the shaft.

State any assumptions made in arriving at your results.   (*UL*)
Answer   2.9 mm; 1.078.

13   A case-hardened shaft is 30 mm diameter with a case depth of 2 mm.

Assuming the case remains perfectly elastic up to its failing stress in shear of 300 MN/m^2 and that the inner core becomes perfectly plastic at a shearing stress of 180 MN/m^2 calculate (a) the torque to cause elastic failure in torsion in the case and (b) the angle of twist per metre length at failure.

State clearly any assumptions made and prove any formula required to deal with plastic conditions. $G = 80$ GN/m^2 for all the material while elastic.   (*UL*)
Answer   1.868 kN m; $\frac{1}{4}$ rad.

14   A hollow cylinder has a bore diameter one half of its outside diameter. The cylinder is made from a material whose stress–strain behaviour is initially elastic and subsequently perfectly plastic. Show that the ratio of axially-applied torques $T_1 : T_2 : T_3$ is equal to $1 : 1 \cdot 18 : 1 \cdot 24$ where

T_1 = the maximum torque to which the cylinder may be subjected without causing plastic deformation,

T_2 = the torque that causes plastic flow to a depth of one-half the thickness of the cylinder wall,

T_3 = the minimum torque to cause plastic flow throughout the entire section of the cylinder.

Assume that the shear strain is directly proportional to the radius.   (*UL*)

15   A simply-supported T-section beam rests as shown between two supports which are 8 metres apart. The beam carries a vertical point load of $W = 12$ kN which acts at mid-span and may be taken to act through the centroid of the beam section. The dimensions of the beam are shown in Figure 9.16 (in mm).

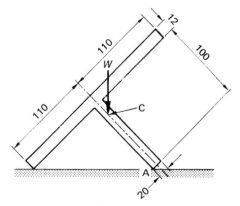

Figure 9.16

Working from first principles, derive the equations for un-symmetrical bending, and hence or otherwise find the maximum bending stress at point A in the beam section, and the position of the neutral axis. *(Brunel)*

Answer 280.0 MN/m²; 99.7° clockwise from axis of symmetry.

16 Figure 9.17 shows the position of the centroid of a 12 cm × 9 cm × 1 cm angle. Determine the positions of the principal axes of the section and the values of the principal second moments of area of the section given that $I_{xx} = 290.0$ cm⁴, $I_{yy} = 140.9$ cm⁴ and the product $I_{xy} = 118.8$ cm⁴.

A length of the angle is used as a horizontal beam simply supported at the ends, to carry a downward load in the plane YY. If the section of the beam is arranged as shown in the figure, find the position of the neutral axis.

Assume throughout that all corners are left square as shown in the figure. *(UL)*

Answer 355.7 cm⁴ at 28° 55′ clockwise from XX; 75.2 cm⁴ at 28° 55′ clockwise from YY; 67° 44′ anticlockwise from XX.

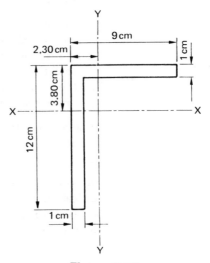

Figure 9.17

17 A 80 mm × 8 mm × 8 mm angle is used as a beam simply supported at each end over a span of 2 m with one leg of the section horizontal and the other vertically upwards.

It is loaded at the centre of the span with a vertical load which may be assumed to pass through the centroid of the section.

The principal second moments of area for the section are 115 cm⁴ and 29.8 cm⁴. The distance of the centroid from the outside edge is 2.26 cm and the toe has a radius of 0.53 cm.

Find the position of the neutral axis and calculate the safe load if the maximum stress is not to exceed 120 MN/m². Graphical constructions may be used. *(UL)*

Hint The toe radius is given since the point of maximum distance from the VV axis lies on the arc of the toe.

Answer 30° 30′ anticlockwise from horizontal; 2.56 kN.

18 The angle shown in Figure 9.18 is used as a beam to span 3 m and to carry a load of 10 kN at mid-span, the load being applied along the direction YY. The second moments of area are $I_{xx} = 492$ cm^4 and $I_{yy} = 172$ cm^4. Find the product of inertia and hence the principal axes and the principal second moments of area of the section.

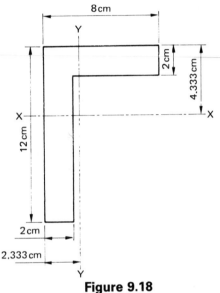

Figure 9.18

Calculate the neutral axis for the given loading and determine the magnitude of the greatest tensile and compressive stresses.

A construction by the Mohr circle is permissible.   (*UL*)

Answer 160 cm^4; principal axes are inclined $22\frac{1}{2}°$ clockwise, to XX and YY; principal second moments are 558.2 and 105.8 cm^4; neutral axis is 42.9 anticlockwise from XX; 161 MN/m^2 (tensile), 142 MN/m^2 (compressive).

19 A section of a beam, unsymmetrical as shown in Figure 9.19

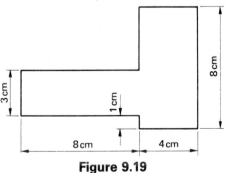

Figure 9.19

is subjected to a bending moment of 2.5 kN m acting in the horizontal plane through the centroid of the section. Determine the magnitude of the stress produced at the bottom right-hand corner. (*UL*)

Answer 52.8 MN/m².

20 Obtain an expression for the central deflection of the plate in Example 9.7. Determine also the maximum bending stress at the centre and the central deflection due to a central load *W* using similar assumptions.

Answer $\dfrac{wL^4B^4}{32Et^3(L^4+B^4)}$; $\dfrac{3WBL^3}{4t^2(L^4+B^4)}$; $\dfrac{WL^3B^3}{16Et^3(L^4+B^4)}$

21 The equation giving the deflected shape of a circular plate of uniform thickness *t* under loading which is normal to the plane containing the rim and symmetrical with respect to its centre, is

$$\frac{d}{dr}\left[\frac{1}{r}\frac{d}{dr}(r\phi)\right]=\frac{F}{D}$$

where ϕ is the slope of any radial line at radius *r*, *F* is the corresponding shearing force per unit length of arc, and *D* is the 'flexural stiffness',

$$Et^3/12(1-v^2)$$

Explain briefly the main assumptions made in deriving this equation.

A diaphragm of light alloy is 15 cm diameter, 1 mm thick and firmly clamped at its edges before loading. Calculate and plot a curve giving the deflected shape of the diaphragm under a uniform pressure of 15 kN/m². Take $E = 70$ GN/m² and $v = 0.3$. (*UL*)

Answer

r (cm)	0	1.5	3	4.5	6	7.5
deflection (mm)	1.157	1.006	0.683	0.340	0.091	0

22 Obtain an expression for the maximum deflection of a circular plate clamped at its outer radius *R* and subjected to a uniform pressure *p* on the whole surface.

Compare this deflection with the maximum deflection of a beam having a rectangular section of unit width and thickness *t*, carrying a load *p* per unit length on a span 2*R* and having its ends fixed.

Answer $\dfrac{3pR^4}{16Et^3}(1-v^2)$; $\dfrac{\text{plate deflection}}{\text{beam deflection}}=\tfrac{3}{8}(1-v^2)$

23 A thin uniform circular flat plate of radius *R* is freely supported at its periphery and loaded in the centre with a load *P* which may be assumed as uniformly distributed over a small circular area of radius r_0.

The following equations apply to this case:

$$\text{(i)} \quad \theta/r = A + \frac{B}{r^2} - \frac{kr^2}{4r_0^2} \left.\begin{array}{l} \\ \\ \\ \\ \end{array}\right\} \quad \text{when} \quad 0 < r < r_0$$

$$\text{(ii)} \quad d\theta/dr = A - \frac{B}{r^2} - \frac{3kr^2}{4r_0^2}$$

$$\text{(iii)} \quad \theta/r = C + \frac{D}{r^2} - k \ln r \left.\begin{array}{l} \\ \\ \\ \\ \end{array}\right\} \quad \text{when} \quad r_0 < r < R$$

$$\text{(iv)} \quad d\theta/dr = C - \frac{D}{r^2} - k(\ln r + 1)$$

where θ is the slope at radius r; A, B, C and D are constants depending on boundary conditions and

$$k = 3(1 - v^2)P/\pi Et^3$$

where v is Poisson's ratio, E is Young's modulus and t is the thickness.

Evaluate the constants A, B, C and D and hence (say by using equation (iii)) deduce that as $r_0 \to 0$ the central deflection tends to

$$\frac{3(1 - v)(3 + v)}{4\pi Et^3} PR^2 \tag{UL}$$

Answer $\quad A = k\left[\ln\dfrac{R}{r_0} + \dfrac{1}{1 + v} - \dfrac{r_0^2}{4R^2}\left(\dfrac{1 - v}{1 + v}\right)\right]; \quad B = 0;$

$\quad C = k\left[\ln R + \dfrac{1}{1 + v} - \dfrac{r_0^2}{4R^2}\left(\dfrac{1 - v}{1 + v}\right)\right]; \quad D = -\tfrac{1}{4}kr_0^2$

24 A pressure vessel is fitted with a circular manhole of 0.6 m diameter, the cover of which is made of a plate 25 mm thick. Assuming that the cover is rigidly clamped around the edge of the manhole, determine the maximum allowable pressure in the vessel if the maximum principal strain in the cover plate must not exceed that produced by a simple direct stress of 150 MN/m². Poisson's ratio, v, for the material of the cover is 0.3.

At any radius r, the bending moment per unit length of arc is

$$M_1 = D\left(\frac{d\theta}{dr} + v\frac{\theta}{r}\right)$$

and the bending moment per unit length of radius is:

$$M_2 = D\left(\frac{\theta}{r} + v\frac{d\theta}{dr}\right)$$

where θ is the slope at radius r and

$$\frac{d}{dr}\left[\frac{1}{r}\frac{d(r\theta)}{dr}\right] + \frac{S}{D} = 0$$

where S = shearing force per unit length of arc at radius r, and the flexural rigidity, D, of the plate is $Et^3/[12(1-v^2)]$.   (UL)
Answer 381.6 kN/m².

25 Dry sand of bulk density ρ is poured on to a uniform disc of radius a, forming a right conical mound of base radius a and height $a/2$. The disc is of thickness t and is fully built-in round its periphery. Young's modulus for the disc material is E and Poisson's ratio is v. It may be assumed that the deflection of the disc does not affect the distribution of load.
   Given that

$$\frac{d}{dr}\left[\frac{1}{r}\frac{d}{dr}(r\phi)\right] = \pm\frac{Q}{D}$$

where Q = shear force/circumferential length at radius r (the sign depends on the convention adopted),
   D = flexural rigidity = $Et^3/12(1-v^2)$
   ϕ = slope of the surface of the disc in the radial direction

show that

   (i) $Q = (3ar - 2r^2)\rho g/12$
   (ii) the maximum deflection of the disc is of the form $\alpha\rho g a^5/D$ and state the value of the constant α, and
   (iii) the maximum radial stress at the edge of the disc is of the form $\beta\rho g a^3/t^2$, and state the value of the constant β.   (UL)

Answer (ii) $\alpha = 43/9\,600$; (iii) $\beta = 29/120$.

26 A circular disc of uniform thickness t is clamped around its periphery and also to a central plunger as shown in Figure 9.20. The effective inner and outer radii are 2.5 cm and 7.5 cm respectively and $t = 1.5$ mm. The plunger exerts a maximum axial force P.
   At any radius r, the general relationship between the slope θ

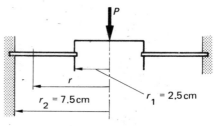

Figure 9.20

and the radius is given by

$$\frac{d}{dr}\left[\frac{1}{r}\frac{d(r\theta)}{dr}\right] + \frac{S}{D} = 0$$

where S is the shearing force per unit length of arc at radius r and

$$D = Et^3/12(1 - v^2)$$

Show that the deflection y at radius r is given by

$$y = -\frac{Pr^2}{8\pi D}(\ln r - 1) + \tfrac{1}{4}C_1 r^2 + C_2 \ln r + C_3$$

and find the maximum deflection for $P = 500\,\text{N}$, assuming $E = 200\,\text{GN/m}^2$ and Poisson's ratio $v = 0.28$.   (*UL*)
Answer 0.97 mm.

27 A curved beam of rectangular section, initially unstressed, is subjected to a bending moment of 1.5 kN m which tends to straighten the bar. The section is 4 cm wide by 5 cm deep in the plane of bending and the mean radius of curvature is 10 cm. Find the position of the neutral axis and the magnitudes of the greatest bending stresses and draw a diagram to show approximately how the stress varies across the section.   (*UL*)
Answer 2.04 mm from central axis; 112.6 and 79.5 MN/m².

28 A curved beam has a rectangular section; the depth of the beam is 5 cm and the radius to which it is curved is 12.5 cm at the mean depth. The beam is subjected at a certain section to a bending moment which would produce a maximum bending stress of 90 MN/m² if the beam had been straight. Determine the percentage increase in the maximum stress due to the curvature.
          (*UL*)

Answer 18.4 per cent.

29 A curved beam has a rectangular section ABCD the depth AB being 9 cm and the width BC 3 cm as shown in Figure 9.21. The radius of curvature at the mean depth is $19\tfrac{1}{2}$ cm, BC being the side where the radius is 15 cm. There is a normal force P acting on the section producing a uniform compressive stress on ABCD and a bending moment M producing compressive stress at BC and tensile stress at AD. The stress at BC was found to be 75 MN/m² compression and that at AD 15 MN/m² tension.

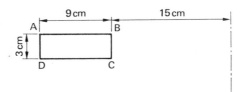

Figure 9.21

Determine the magnitudes of P and M. Work from first principles. (UL)

Answer   142.8 kN; 8.35 kN m.

30   Figure 9.22 shows a machine frame and the load to which it is subjected. Determine the stresses acting across the section AB at A and B. Make a diagram showing the distribution of stress across the section. Flange and web thickness 3 cm. (UL)
Answer $\sigma_A = 12.46$ MN/m² (compression); $\sigma_B = 19.73$ MN/m² (tension).

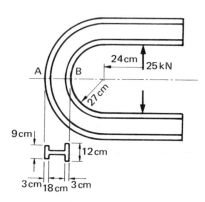

Figure 9.22

31   A bar having a rectangular cross-section 9 cm by 6 cm is bent to a semicircular form, the plane of the semi-circle being that of the 9 cm dimension; the inner and outer radii of curvature are $13\frac{1}{2}$ cm and $22\frac{1}{2}$ cm respectively. The half-ring stands on a frictionless horizontal surface with the plane of the semicircle vertical. A vertical load of 80 kN is applied to the mid-point of the outer semicircle, this point being mid-way between the two points of support which are 36 cm apart.

   Find the maximum tensile and compressive stresses due to bending at a section near the point of application of the load. The depth of the beam is to be considered large compared with the radius of curvature. (UL)
Answer   111.1 MN/m² (tensile); 78.6 MN/m² (compressive).

32   A steel member of rectangular section, breadth $b = 3$ cm and depth $d = 8$ cm is formed to a U-shape and subjected to a diametral load $W = 22$ kN as shown in Figure 9.23. Find the greatest value for the mean radius R if the maximum allowable stress is 120 MN/m². Denoting the distance of the neutral axis for the bending stresses from the centroid as e, you may assume that $e = d^2/12R$.

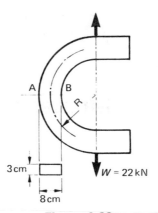

Figure 9.23

Make a diagram showing the distribution of normal stress across the section AB. (*UL*)
Answer $R = 96$ mm; $\sigma_B = 120$ MN/m^2 (tensile); $\sigma_A = 53.08$ MN/m^2 (compressive).

Appendix 1

Moments of area

Product moment

Suppose (Figure A.1) XX and YY are perpendicular axes and δA is an element of area having coordinates (x, y). The second moments of area of the complete figure about these axes are

$$I_{XX} = \int y^2 \, dA \quad \text{and} \quad I_{YY} = \int x^2 \, dA$$

The product moment (or product of inertia) is defined as

$$I_{XY} = \int xy \, dA \tag{i}$$

For axes of symmetry the product moment is zero since x and y follow the usual sign conventions and corresponding parts of the figure on opposite sides of the axes cancel each other.

The theorem of parallel axes used in calculating second moments has its counterpart in product moments. Suppose, Figure A.2, XX and YY are principal axes of a given area which intersect at O and AA and BB are parallel axes distances b and a respectively from them. Then the product moment for axes AA and BB is

$$I_{AB} = \int (x + a)(y + b) \, dA$$

$$= \int xy \, dA + a \int y \, dA + b \int x \, dA + ab \int dA$$

Since O is the centroid and XX, YY are principal axes the first three

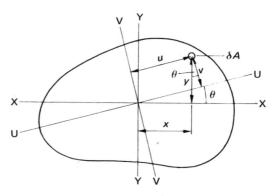

Figure A.1

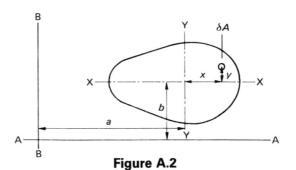

Figure A.2

integrals in this expression are all zero. Thus

$$I_{AB} = ab \times \text{(total area of the figure)}$$

Directions of principal axes

For an unsymmetrical figure the directions of the principal axes can be determined from the condition that the product moment is zero. Let O (Figure A.1) be the centroid of the figure, XX and YY perpendicular axes which are not principal axes. Let the principal axes UU and VV make an angle θ as shown with XX and YY.

From the geometry of Figure A.1 the coordinates of δA relative to UU and VV are

$$u = x \cos \theta + y \sin \theta$$
$$v = y \cos \theta - x \sin \theta$$

(ii)

The product moment for UU and VV is therefore

$$I_{UV} = \int uv \, dA$$

$$= \int (x \cos \theta + y \sin \theta)(y \cos \theta - x \sin \theta) \, dA$$

$$= \cos \theta \sin \theta \left[\int y^2 \, dA - \int x^2 \, dA \right] + (\cos^2 \theta - \sin^2 \theta) \int xy \, dA$$

$$= \tfrac{1}{2} \sin 2\theta (I_{XX} - I_{YY}) + \cos 2\theta \, I_{XY}$$

For UU and VV to be principal axes, $I_{UV} = 0$ and, thus, equating the last result to zero,

$$\tan 2\theta = \frac{2I_{XY}}{I_{YY} - I_{XX}}$$

(iii)

Hence the principal second moments are, using (ii),

$$I_{UU} = \int v^2 \, dA$$

$$= \int (y \cos \theta - x \sin \theta)^2 \, dA$$

$$= \cos^2 \theta \int y^2 \, dA + \sin^2 \theta \int x^2 \, dA - 2 \sin \theta \cos \theta \int xy \, dA$$

$$= I_{XX} \cos^2 \theta + I_{YY} \sin^2 \theta - I_{XY} \sin 2\theta$$

(iv)

and similarly

$$I_{VV} = I_{XX} \sin^2 \theta + I_{YY} \cos^2 \theta + I_{XY} \sin 2\theta \qquad \text{(v)}$$

Adding (iv) and (v)

$$I_{UU} + I_{VV} = I_{XX} + I_{YY} \qquad \text{(vi)}$$

a result which can also be obtained by noting that the sum of the second moments for a pair of perpendicular axes is equal to the polar second moment J, since $r^2 = x^2 + y^2 = u^2 + v^2$.

Using the identities $\cos^2 \theta = \frac{1}{2}(\cos 2\theta + 1)$ and $\sin^2 \theta = \frac{1}{2}(1 - \cos 2\theta)$ together with (iii), equation (iv) becomes

$$I_{UU} = \tfrac{1}{2}(I_{XX} + I_{YY}) + \tfrac{1}{2}(I_{XX} - I_{YY}) \cos 2\theta - I_{XY} \sin 2\theta$$

$$= \tfrac{1}{2}(I_{XX} + I_{YY}) - I_{XY} \frac{\cos 2\theta}{\tan 2\theta} - I_{XY} \sin 2\theta$$

$$= \tfrac{1}{2}(I_{XX} + I_{YY}) - I_{XY}\left(\frac{\cos^2 2\theta + \sin^2 2\theta}{\sin 2\theta}\right)$$

$$= \tfrac{1}{2}(I_{XX} + I_{YY}) - I_{XY} \operatorname{cosec} 2\theta \qquad \text{(vii)}$$

similarly from (v)

$$I_{VV} = \tfrac{1}{2}(I_{XX} + I_{YY}) + I_{XY} \operatorname{cosec} 2\theta \qquad \text{(viii)}$$

Multiplying these results together, noting that $\operatorname{cosec}^2 2\theta = 1 + \cot^2 2\theta$, and again using (viii), we have

$$I_{UU} I_{VV} = \tfrac{1}{4}(I_{XX} + I_{YY})^2 - I_{XY}^2 \operatorname{cosec}^2 2\theta$$

$$= \tfrac{1}{4}(I_{XX} + I_{YY})^2 - I_{XY}^2 (1 + \cot^2 2\theta)$$

$$= \tfrac{1}{4}(I_{XX} + I_{YY})^2 - I_{XY}^2\left(\frac{I_{YY} - I_{XX}}{2 I_{XY}}\right)^2 - I_{XY}^2$$

$$= \tfrac{1}{4}[(I_{XX} + I_{YY})^2 - (I_{YY} - I_{XX})^2] - I_{XY}^2$$

$$= I_{XX} I_{YY} - I_{XY}^2 \qquad \text{(ix)}$$

Subtracting (vii) from (viii)

$$I_{VV} - I_{UU} = 2 I_{XY} \operatorname{cosec} 2\theta \qquad \text{(x)}$$

Application of Mohr circle

The form of equations (vii) and (viii) suggests that the Mohr circle (see Chapters 1 and 2) can be used to determine I_{UU} and I_{VV}. If I_{XX} and I_{YY} are set out along the horizontal axis, Figure A.3, and I_{XY} is drawn as an ordinate at each point then the circle passing through P_1 and P_2 intersects the horizontal axis at points which represent I_{UU} and I_{VV}. The radius of the circle is $I_{XY} \operatorname{cosec} 2\theta$ and the details of the proof are left to the reader.

The inclination of the principal axes to XX and YY can also be determined by measuring θ on the diagram.

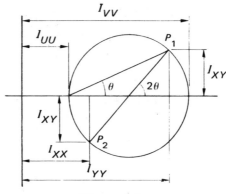

Figure A.3

Formulae for principal second moments

By analogy with the relationships for principal stresses (see equation (iii) on page 14) the principal second moments I_{UU} and I_{VV} are the roots of the following quadratic equation for I

$$(I - I_{XX})(I - I_{YY}) = I_{XY}{}^2$$

from which

$$I_{UU} \quad \text{and} \quad I_{VV} = \tfrac{1}{2}(I_{XX} + I_{YY}) \pm \tfrac{1}{2}\sqrt{[(I_{XX} - I_{YY})^2 + 4I_{XY}{}^2]}$$

The same results can be obtained from the geometry of Figure A.3 or analytically, by eliminating θ from equations (iii), (vii) and (viii) above.

Appendix 2

Units

All the problems and solutions in this book use SI (Système Internationale) units. The system is based on the fundamental units: *kilogram* (kg) for mass, *metre* (m) for length and *second* (s) for time. The paragraphs that follow describe the units of other physical quantities that occur frequently in solid mechanics and show how errors can be avoided when very large and very small values occur in numerical examples. Other systems of units may be encountered and a table of factors is provided to enable conversions to be made to and from SI.

Force and stress

Force is measured in *newtons* (symbol N), and stress, which is the ratio force/area, therefore has the units *newtons per square metre* (N/m^2). The name *pascal* (Pa) has been given to this unit and it has been adopted by some authorities. We have used N/m^2 as it has the advantage of emphasizing the physical relationship between stress, force and area.

A complication arises in numerical examples since the stress values usually exceed $10^6 N/m^2$. It is therefore convenient to express the values of stress in MN/m^2 or N/mm^2. Numerical values will be the same in the two cases but we have used the form MN/m^2 throughout. Elastic moduli have the same basic units as stress but practical values are even higher and exceed $10^9 N/m^2$ for many materials. In this book the form GN/m^2 has been used but the alternative kN/mm^2 will be found in some examination papers. Numerical values are the same in the two cases.

Second moment of area

A special problem arises with the quantity *second moment of area*. Its numerical value in m^4 is usually very small but in mm^4 it becomes very large. As a result it is often expressed in cm^4 and it should be noted that $1 cm^4 = 10^{-8} m^4$.

Table of SI units

Quantity	Derivation	Name of unit	Symbol
Mass	—	kilogram	kg
Length	—	metre	m
Time	—	second	s
Area	(length)^2	—	m^2

Table of SI units (continued)

Quantity	Derivation	Name of unit	Symbol
Volume	$(\text{length})^3$	—	m^3
First moment of area	area × length	—	m^3
Second moment of area	area × $(\text{length})^2$	—	m^4
Velocity	length ÷ time	—	m/s
Acceleration	velocity ÷ time	—	m/s^2
Force	mass × acceleration	newton	$N\ (=kg\,m/s^2)$
Stress and pressure	force ÷ area	(pascal)	N/m^2 (or Pa)
Torque and moment	force × distance	—	N m
Work and energy	force × distance	joule	$J\ (=N\,m)$
Power	work ÷ time	watt	$W\ (=N\,m/s)$

Multiples and sub-multiples

Factor	Unit prefix	Symbol
$1\ 000\ 000\ 000 = 10^9$	giga-	G
$1\ 000\ 000 = 10^6$	mega-	M
$1\ 000 = 10^3$	kilo-	k
$0.01 = \dfrac{1}{10^2} = 10^{-2}$	centi-	c
$0.001 = \dfrac{1}{10^3} = 10^{-3}$	milli-	m
$0.000\ 001 = \dfrac{1}{10^6} = 10^{-6}$	micro-	μ

Problems using multiple units

We recommend that, in tackling numerical problems, the prefixes to units are replaced by factors that are powers of 10. At the end of the working the prefix form can be adopted for the answer.

Example. The central deflection δ of a centrally-loaded uniform beam is given by $\delta = WL^3/48EI$. Calculate δ when $W = 144$ kN, $L = 4$ m, $E = 200$ GN/m^2 and $I = 8\,000$ cm^4. The work may be set out as follows:

$$\delta = \frac{144\ \text{kN} \times (4\ \text{m})^3}{48 \times 200\ \text{GN/m}^2 \times 8\,000\ \text{cm}^4}$$

$$= \frac{(144 \times 10^3\ \text{N}) \times (4\ \text{m})^3}{48 \times (200 \times 10^9\ \text{N/m}^2) \times (8\,000 \times 10^{-8}\ \text{m}^4)}$$

$$= \frac{144 \times 4^3}{48 \times 200 \times 8\,000} \times \frac{10^3}{10^9 \times 10^{-8}} \frac{\text{N} \times \text{m}^3}{(\text{N/m}^2) \times \text{m}^4}$$

$$= \frac{12}{100 \times 1\,000} \times \frac{10^3 \times 10^8}{10^9}\ \text{m}$$

$$= 12 \times 10^{-3}\ \text{m}$$

$$= 12\ \text{mm}$$

This layout may seem elaborate but it helps to avoid errors and it can be shortened as confidence in the method is gained.

Conversion of units

Some examinations and textbooks contain examples in imperial or metric gravitational units. The table below can be used in these cases to convert numerical values to and from SI units. The conversion factors have been rounded to four significant figures and will be sufficiently accurate for most purposes. They are derived from the exact basic values, $1\,\text{ft} = 0.304\,8\,\text{m}$ and $1\,\text{lb} = 0.453\,592\,37\,\text{kg}$.

Quantity	To SI units		From SI units	
Length	1 in	$= 25.4\,\text{mm}$	1 mm	$= 0.039\,37\,\text{in}$
	1 ft	$= 0.304\,8\,\text{m}$	1 m	$= 3.281\,\text{ft}$
Area	$1\,\text{in}^2$	$= 6.452\,\text{cm}^2$	$1\,\text{cm}^2$	$= 0.155\,\text{in}^2$
	$1\,\text{in}^2$	$= 645.2\,\text{mm}^2$	$1\,\text{mm}^2$	$= 0.001\,55\,\text{in}^2$
	$1\,\text{ft}^2$	$= 0.092\,90\,\text{m}^2$	$1\,\text{m}^2$	$= 10.76\,\text{ft}^2$
Volume and section modulus	$1\,\text{in}^3$	$= 16.39\,\text{cm}^3$	$1\,\text{cm}^3$	$= 0.061\,02\,\text{in}^3$
Second moment of area	$1\,\text{in}^4$	$= 41.62\,\text{cm}^4$	$1\,\text{cm}^4$	$= 0.024\,03\,\text{in}^4$
Mass	1 lb	$= 0.453\,6\,\text{kg}$	1 kg	$= 2.205\,\text{lb}$
Mass per unit length	1 lb/ft	$= 1.488\,\text{kg/m}$	1 kg/m	$= 0.672\,0\,\text{lb/ft}$
Force or load	1 lbf	$= 4.448\,\text{N}$	1 N	$= 0.224\,8\,\text{lbf}$
	1 tonf	$= 9.964\,\text{kN}$	1 kN	$= 0.100\,4\,\text{tonf}$
	1 kgf	$= 9.807\,\text{N}$	1 N	$= 0.102\,0\,\text{kgf}$
	1 tonnef	$= 9.807\,\text{kN}$	1 kN	$= 0.102\,0\,\text{tonnef}$
	1 kip	$= 4.448\,\text{kN}$	1 kN	$= 0.224\,8\,\text{kip}$
Load per unit length	1 lbf/ft	$= 14.59\,\text{N/m}$	1 N/m	$= 0.068\,52\,\text{lbf/ft}$
	1 tonf/ft	$= 32.69\,\text{kN/m}$	1 kN/m	$= 0.030\,59\,\text{tonf/ft}$
	1 kgf/m	$= 9.807\,\text{N/m}$	1 N/m	$= 0.102\,0\,\text{kgf/m}$
	1 tonnef/m	$= 9.807\,\text{kN/m}$	1 kN/m	$= 0.102\,0\,\text{tonnef/m}$
	1 kip/ft	$= 14.59\,\text{kN/m}$	1 kN/m	$= 0.068\,52\,\text{kip/ft}$
Moment or torque	1 lbf ft	$= 1.356\,\text{N m}$	1 N m	$= 0.737\,6\,\text{lbf ft}$
	1 tonf ft	$= 3.037\,\text{kN m}$	1 kN m	$= 0.329\,3\,\text{tonf ft}$
	1 kgf m	$= 9.807\,\text{N m}$	1 N m	$= 0.102\,0\,\text{kgf m}$
	1 tonnef m	$= 9.807\,\text{kN m}$	1 kN m	$= 0.102\,0\,\text{tonnef m}$
	1 kip ft	$= 1.356\,\text{kN m}$	1 kN m	$= 0.737\,6\,\text{kip ft}$
Work and energy	1 ft lbf	$= 1.356\,\text{J}$	1 J	$= 0.737\,6\,\text{ft lbf}$
	1 ft tonf	$= 3.037\,\text{kJ}$	1 kJ	$= 0.329\,3\,\text{ft tonf}$
	1 m kgf	$= 9.807\,\text{J}$	1 J	$= 0.102\,0\,\text{m kgf}$
Power	1 hp	$= 0.745\,7\,\text{kW}$	1 kW	$= 1.341\,\text{hp}$
	1 metric hp	$= 0.735\,5\,\text{kW}$	1 kW	$= 1.360\,\text{metric hp}$
Stress and pressure	$1\,\text{lbf/in}^2$	$= 6.895\,\text{kN/m}^2$	$1\,\text{kN/m}^2$	$= 0.145\,0\,\text{lbf/m}^2$
	$1\,\text{tonf/in}^2$	$= 15.44\,\text{MN/m}^2$	$1\,\text{MN/m}^2$	$= 0.064\,75\,\text{tonf/in}^2$
	$1\,\text{kgf/m}^2$	$= 9.807\,\text{N/m}^2$	$1\,\text{N/m}^2$	$= 0.102\,0\,\text{kgf/m}^2$
	$1\,\text{tonnef/m}^2$	$= 9.807\,\text{kN/m}^2$	$1\,\text{kN/m}^2$	$= 0.102\,0\,\text{tonnef/m}^2$
	$1\,\text{kip/in}^2$	$= 6.895\,\text{MN/m}^2$	$1\,\text{MN/m}^2$	$= 0.145\,0\,\text{kip/in}^2$

Index